Probability with Applications in Science and Engineering

Probability with Applications in Science and Engineering

Edited by
Michael Fuller

Willford Press
www.willfordpress.com

Published by Willford Press,
118-35 Queens Blvd., Suite 400,
Forest Hills, NY 11375, USA

ISBN: 978-1-68285-486-0

Cataloging-in-Publication Data

Probability with applications in science and engineering / edited by Michael Fuller.
p. cm.
Includes bibliographical references and index.
ISBN 978-1-68285-486-0
1. Probabilities. 2. Science--Statistical methods. 3. Engineering--Statistical methods.
I. Fuller, Michael.
QA273 .P76 2018
519.2--dc23

For information on all Willford Press publications
visit our website at www.willfordpress.com

WILLFORD PRESS

Printed in China.

Contents

Preface

The vast fields of science and engineering are closely related because engineering uses science to create the new technology, which in turn helps the subject to grow and discover new scientific facts. Engineering employs elements of mathematics, social knowledge, science and economics to design, invent, create, build and maintain best scientific solutions to problems. Science then uses these machines, hardware, software, etc. to discover and prove new facts and truths about the universe. The various sub-fields of science and technology with respect to probability, along with technological progress that have future implications are glanced at in this book. It includes different approaches, evaluations and methodologies on the applications of probability in science and engineering. This textbook attempts to assist those with a goal of delving into this field.

Given below is the chapter wise description of the book:

Chapter 1- Uncertainties relating to processes, artificial or natural, call for the study of statistics and probability. These uncertainties may be caused by lack of adequate knowledge about natural processes, inherent variability, or inappropriately chosen model to study a process. In civil engineering, the data are random and inculcate many variables, which necessitate the application of statistics and probability. This chapter will provide an integrated understanding of probability methods in civil engineering.

Chapter 2- Probability is the study of chance and events that may take place or not. It calculates the possibility of an event to happen, with the results falling between zero and one. The closer the evaluation of an event is to one, the higher is its chances to occur. This chapter elucidates the crucial theories and principles of probability.

Chapter 3- Random variables measure the probability of a random phenomenon. While the process requires dealing with real numbers and accurate probabilities, the logic behind constituting the phenomenon itself is random or not entirely known. The aspects elucidated in this chapter are of vital importance, and provide a better understanding of random variables.

Chapter 4- Probability distribution is mathematical function that provides the set of all the possible outcomes of an experiment. It can be divided into two parts, namely discrete probability distribution and continuous probability distribution. While the former deals with situations which have a finite number of outcomes, the latter is related to scenarios that have an infinite number of possibilities. The major components of probability distribution are discussed in this chapter.

Chapter 5- The population data is evaluated following probability distribution and verified through samples. The section explores several of these methods. In simple random sample, a subset is chosen from a larger set in a way such that all the elements have the same probability of getting selected. This chapter has been carefully written to provide an easy understanding of the varied facets of sample distribution.

Chapter 6- Statistical hypothesis testing helps in finding confirmatory intervals after estimates from sample data are acquired. In this method, two data, normally including a sample data, are compared to approve or disapprove a hypothesis. The chapter delves into the concept of regression analysis as well. All the diverse principles of statistical hypothesis testing have been carefully analyzed in this chapter.

Indeed, my job was extremely crucial and challenging as I had to ensure that every chapter is informative and structured in a student-friendly manner. I am thankful for the support provided by my family and colleagues during the completion of this book.

Editor

1

Basics of Probability Methods in Civil Engineering

Uncertainties relating to processes, artificial or natural, call for the study of statistics and probability. These uncertainties may be caused by lack of adequate knowledge about natural processes, inherent variability, or inappropriately chosen model to study a process. In civil engineering, the data are random and inculcate many variables, which necessitate the application of statistics and probability. This chapter will provide an integrated understanding of probability methods in civil engineering.

Variability in the Data

Uncertainties about certain pure natural processes as well as man-made experiments that are involving some natural elements necessitate the study of probability and statistics. These uncertainties may result in due to,

- the inherent variability (also known as Aleatory variability),
- lack of complete knowledge (Epistemic uncertainty) about the natural process, and
- the inappropriate model chosen to study the process

Aleatory Variability

In many of the civil engineering fields the data are random in nature and contain lot of variability and the reasons for the variability are beyond the comprehension of the civil engineer. This inherent variability is also known as the Aleatory variability. The best example is the rainfall in a particular month of the year and the variability in the magnitude of the rainfall, when observed over several years, is completely random. Similarly, the time gaps between the vehicle arrivals (during off-peak hours) at a particular location of the road are completely random and the gap data between the vehicular arrivals contain lot of variability.

Epistemic Uncertainty

Unlike the Aleatory variability, the epistemic uncertainty results in due to the inability of the modeler to comprehend the system in the form of a mathematical/simulation model.

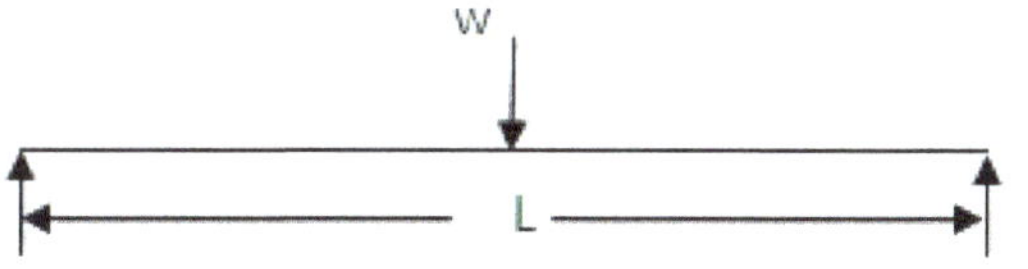

Simply supported beam with concentrated load

Deflection of a simply supported beam under a concentrated load, can be found easily using the following formula.

$$y_c = \frac{wL^3}{48EI}$$

Where L is the length of the beam, W is the load applied, E is the young's modulus, and *I* is the moment of Inertia of the beam section.

But this formula is based on several assumptions like the ones given below;

1. The material is linearly elastic
2. Even after bending the plane section of the beam remains plane
3. Stress strain relationship is same in compression and tension

Due to these assumptions the actual deflection and the deflection value obtained from the above formula are different. When the same experiment is repeated under exactly similar conditions estimated deflections may vary. This variability is resulting due to the assumptions made in deriving the deflection formula.

Modeling the Outcome of a Die

When a fair die is rolled the probability of getting any number between 1 and 6 is 1/6 and it is same for all the numbers. It is also difficult to say what would be the probable outcome when the die is rolled. This uncertainty is very difficult to explain and may be termed as aleatory uncertainty.

Suppose a die is rolled for six times, the probabilities associated to the each outcome may be as follows. As discussed earlier, the probability associated to each outcome in case, if the die is fair, 1/6, but if it is an unfair die the associated probabilities are to be calculated based on the outcomes of the six rolls. If the six outcomes are 1, 2, 2, 3, 4, and 5 the corresponding probabilities may be as given in the following table.

Table: The variability associated to the outcomes of an unknown die

Outcome	Probability, Fair die	Probability, Unfair die
1	1/6	1/6
2	1/6	2/6
3	1/6	1/6
4	1/6	1/6
5	1/6	1/6
6	1/6	0

It is possible to get an idea about the kind of die that is being rolled when the experiment is repeated for several times. If the die is fair the probabilities tend to 1/6 and if it is unfair the probability associated to a particular outcome tends to be zero. This kind of uncertainty may be termed as epistemic uncertainty.

The term 'randomness' is used for inherent variability and the term 'uncertainty' is used for lack of complete knowledge. It is not easy to differentiate randomness and uncertainty and it depends on the level of understanding about the process. When the experiments involving some natural elements are repeated under similar conditions the outcomes may or may not be the same. Quantification of these uncertainties and the evaluation of their effects on design and implementation require concepts and methods of probability.

Civil Engineering

The Burj Khalifa, the world's tallest skyscraper, 828 metres (2,717 ft) tall

A multi-level stack interchange, buildings, houses, and park in Shanghai, China.

Philadelphia City Hall in the US is still the world's tallest masonry load-bearing structure.

Civil engineering is a professional engineering discipline that deals with the design, construction, and maintenance of the physical and naturally built environment, including works like roads, bridges, canals, dams, and buildings. Civil engineering is traditionally broken into a number of sub-disciplines. It is the second-oldest engineering discipline after military engineering, and it is defined to distinguish non-military engineering from military engineering. Civil engineering takes place in the public sector from municipal through to national governments, and in the private sector from individual homeowners through to international companies.

History of the Civil Engineering Profession

Leonhard Euler developed the theory explaining the buckling of columns

Engineering has been an aspect of life since the beginnings of human existence. The earliest practice of civil engineering may have commenced between 4000 and 2000 BC in ancient Egypt, the

Indus Valley Civilization, and Mesopotamia (ancient Iraq) when humans started to abandon a nomadic existence, creating a need for the construction of shelter. During this time, transportation became increasingly important leading to the development of the wheel and sailing.

Until modern times there was no clear distinction between civil engineering and architecture, and the term engineer and architect were mainly geographical variations referring to the same occupation, and often used interchangeably. The construction of pyramids in Egypt (circa 2700–2500 BC) were some of the first instances of large structure constructions. Other ancient historic civil engineering constructions include the Qanat water management system (the oldest is older than 3000 years and longer than 71 km,) the Parthenon by Iktinos in Ancient Greece (447–438 BC), the Appian Way by Roman engineers (c. 312 BC), the Great Wall of China by General Meng T'ien under orders from Ch'in Emperor Shih Huang Ti (c. 220 BC) and the stupas constructed in ancient Sri Lanka like the Jetavanaramaya and the extensive irrigation works in Anuradhapura. The Romans developed civil structures throughout their empire, including especially aqueducts, insulae, harbors, bridges, dams and roads.

John Smeaton, the "father of civil engineering"

In the 18th century, the term civil engineering was coined to incorporate all things civilian as opposed to military engineering. The first self-proclaimed civil engineer was John Smeaton, who constructed the Eddystone Lighthouse. In 1771 Smeaton and some of his colleagues formed the Smeatonian Society of Civil Engineers, a group of leaders of the profession who met informally over dinner. Though there was evidence of some technical meetings, it was little more than a social society.

In 1818 the Institution of Civil Engineers was founded in London, and in 1820 the eminent engineer Thomas Telford became its first president. The institution received a Royal Charter in 1828, formally recognising civil engineering as a profession. Its charter defined civil engineering as:

the art of directing the great sources of power in nature for the use and convenience of man, as the means of production and of traffic in states, both for external and internal trade, as applied in the construction of roads, bridges, aqueducts, canals, river navigation and docks for internal intercourse and exchange, and in the construction of ports, harbours, moles, breakwaters and lighthouses, and in the art of navigation by artificial power for the purposes of commerce, and in the construction and application of machinery, and in the drainage of cities and towns.

History of Civil Engineering Education

The first private college to teach civil engineering in the United States was Norwich University, founded in 1819 by Captain Alden Partridge. The first degree in civil engineering in the United States was awarded by Rensselaer Polytechnic Institute in 1835. The first such degree to be awarded to a woman was granted by Cornell University to Nora Stanton Blatch in 1905.

In the UK during the early 19th century, the division between civil engineering and military engineering (served by the Royal Military Academy, Woolwich), coupled with the demands of the Industrial Revolution, spawned new engineering education initiatives: the Royal Polytechnic Institution was founded in 1838, the private College for Civil Engineers in Putney was established in 1839, and the UK's first Chair of Engineering was established at the University of Glasgow in 1840.

History of Civil Engineering

Chichen Itza was a large pre-Columbian city in Mexico built by the Maya people of the Post Classic. The northeast column temple also covers a channel that funnels all the rainwater from the complex some 40 metres (130 ft) away to a rejollada, a former cenote.

A Roman aqueduct [built circa 19 BC] near Pont du Gard, France

Civil engineering is the application of physical and scientific principles for solving the problems of society, and its history is intricately linked to advances in understanding of physics and mathematics throughout history. Because civil engineering is a wide-ranging profession, including several specialized sub-disciplines, its history is linked to knowledge of structures, materials science, geography, geology, soils, hydrology, environment, mechanics and other fields.

Throughout ancient and medieval history most architectural design and construction was carried out by artisans, such as stonemasons and carpenters, rising to the role of master builder. Knowledge was retained in guilds and seldom supplanted by advances. Structures, roads and infrastructure that existed were repetitive, and increases in scale were incremental.

One of the earliest examples of a scientific approach to physical and mathematical problems applicable to civil engineering is the work of Archimedes in the 3rd century BC, including Archimedes Principle, which underpins our understanding of buoyancy, and practical solutions such as Archimedes' screw. Brahmagupta, an Indian mathematician, used arithmetic in the 7th century AD, based on Hindu-Arabic numerals, for excavation (volume) computations.

The Civil Engineer

Education and Licensure

Surveying students with professor at the Helsinki University of Technology in the late 19th century.

Civil engineers typically possess an academic degree in civil engineering. The length of study is three to five years, and the completed degree is designated as a bachelor of engineering, or a bachelor of science in engineering. The curriculum generally includes classes in physics, mathematics, project management, design and specific topics in civil engineering. After taking basic courses in most sub-disciplines of civil engineering, they move onto specialize in one or more sub-disciplines at advanced levels. While an undergraduate degree (BEng/BSc) normally provides successful students with industry-accredited qualification, some academic institutions offer post-graduate degrees (MEng/MSc), which allow students to further specialize in their particular area of interest.

In most countries, a bachelor's degree in engineering represents the first step towards professional certification, and a professional body certifies the degree program. After completing a certified degree program, the engineer must satisfy a range of requirements (including work experience and exam requirements) before being certified. Once certified, the engineer is designated as a professional engineer (in the United States, Canada and South Africa), a chartered engineer (in most Commonwealth countries), a chartered professional engineer (in Australia and New Zealand), or a European engineer (in most countries of the European Union). There are international agreements between relevant professional bodies to allow engineers to practice across national borders.

The benefits of certification vary depending upon location. For example, in the United States and Canada, "only a licensed professional engineer may prepare, sign and seal, and submit engineering plans and drawings to a public authority for approval, or seal engineering work for public and private clients." This requirement is enforced under provincial law such as the Engineers Act in Quebec.

No such legislation has been enacted in other countries including the United Kingdom. In Australia, state licensing of engineers is limited to the state of Queensland. Almost all certifying bodies maintain a code of ethics which all members must abide by.

Engineers must obey contract law in their contractual relationships with other parties. In cases where an engineer's work fails, he may be subject to the law of tort of negligence, and in extreme cases, criminal charges. An engineer's work must also comply with numerous other rules and regulations such as building codes and environmental law.

Sub-disciplines

The Akashi Kaikyō Bridge in Japan, currently the world's longest suspension span.

In general, civil engineering is concerned with the overall interface of human created fixed projects with the greater world. General civil engineers work closely with surveyors and specialized civil engineers to design grading, drainage, pavement, water supply, sewer service, dams, electric and communications supply. General civil engineering is also referred to as site engineering, a branch of civil engineering that primarily focuses on converting a tract of land from one usage to another. Site engineers spend time visiting project sites, meeting with stakeholders, and preparing construction plans. Civil engineers apply the principles of geotechnical engineering, structural engineering, environmental engineering, transportation engineering and construction engineering to residential, commercial, industrial and public works projects of all sizes and levels of construction.

Materials Science and Engineering

Materials science is closely related to civil engineering. It studies fundamental characteristics of materials, and deals with ceramics such as concrete and mix asphalt concrete, strong metals such as aluminum and steel, and thermosetting polymers including polymethylmethacrylate (PMMA) and carbon fibers.

Materials engineering involves protection and prevention (paints and finishes). Alloying combines two types of metals to produce another metal with desired properties. It incorporates elements of applied physics and chemistry. With recent media attention on nanoscience and nanotechnology, materials engineering has been at the forefront of academic research. It is also an important part of forensic engineering and failure analysis.

Coastal Engineering

Oosterscheldekering, a storm surge barrier in the Netherlands.

Coastal engineering is concerned with managing coastal areas. In some jurisdictions, the terms sea defense and coastal protection mean defense against flooding and erosion, respectively. The term coastal defense is the more traditional term, but coastal management has become more popular as the field has expanded to techniques that allow erosion to claim land.

Construction Engineering

Construction engineering involves planning and execution, transportation of materials, site development based on hydraulic, environmental, structural and geotechnical engineering. As construction firms tend to have higher business risk than other types of civil engineering firms do, construction engineers often engage in more business-like transactions, for example, drafting and reviewing contracts, evaluating logistical operations, and monitoring prices of supplies.

Earthquake Engineering

Earthquake engineering involves designing structures to withstand hazardous earthquake exposures. Earthquake engineering is a sub-discipline of structural engineering. The main objectives of earthquake engineering are to understand interaction of structures on the shaky ground; foresee the consequences of possible earthquakes; and design, construct and maintain structures to perform at earthquake in compliance with building codes.

Environmental Engineering

Water pollution

Environmental engineering is the contemporary term for sanitary engineering, though sanitary engineering traditionally had not included much of the hazardous waste management and environmental remediation work covered by environmental engineering. Public health engineering and environmental health engineering are other terms being used.

Environmental engineering deals with treatment of chemical, biological, or thermal wastes, purification of water and air, and remediation of contaminated sites after waste disposal or accidental contamination. Among the topics covered by environmental engineering are pollutant transport, water purification, waste water treatment, air pollution, solid waste treatment, and hazardous waste management. Environmental engineers administer pollution reduction, green engineering, and industrial ecology. Environmental engineers also compile information on environmental consequences of proposed actions.

Geotechnical Engineering

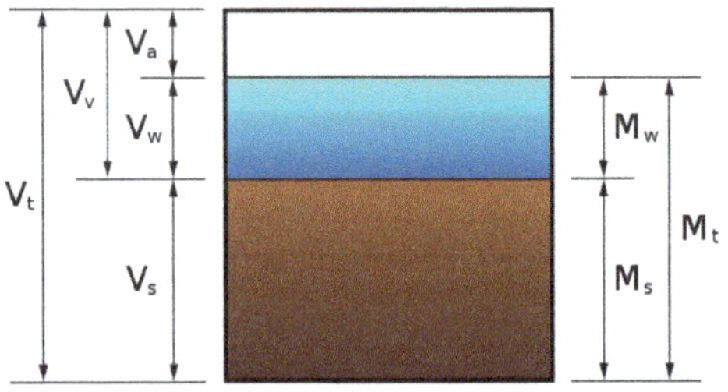

A phase diagram of soil indicating the weights and volumes of air, soil, water, and voids.

Geotechnical engineering studies rock and soil supporting civil engineering systems. Knowledge from the field of soil science, materials science, mechanics, and hydraulics is applied to safely and economically design foundations, retaining walls, and other structures. Environmental efforts to protect groundwater and safely maintain landfills have spawned a new area of research called geo-environmental engineering.

Identification of soil properties presents challenges to geotechnical engineers. Boundary conditions are often well defined in other branches of civil engineering, but unlike steel or concrete, the material properties and behavior of soil are difficult to predict due to its variability and limitation on investigation. Furthermore, soil exhibits nonlinear (stress-dependent) strength, stiffness, and dilatancy (volume change associated with application of shear stress), making studying soil mechanics all the more difficult.

Water Resources Engineering

Hoover Dam

Water resources engineering is concerned with the collection and management of water (as a natural resource). As a discipline it therefore combines elements of hydrology, environmental science, meteorology, conservation, and resource management. This area of civil engineering relates to the prediction and management of both the quality and the quantity of water in both underground (aquifers) and above ground (lakes, rivers, and streams) resources. Water resource engineers analyze and model very small to very large areas of the earth to predict the amount and content of water as it flows into, through, or out of a facility. Although the actual design of the facility may be left to other engineers.

Hydraulic engineering is concerned with the flow and conveyance of fluids, principally water. This area of civil engineering is intimately related to the design of pipelines, water supply network, drainage facilities (including bridges, dams, channels, culverts, levees, storm sewers), and canals. Hydraulic engineers design these facilities using the concepts of fluid pressure, fluid statics, fluid dynamics, and hydraulics, among others.

Structural Engineering

Structural engineering is concerned with the structural design and structural analysis of buildings, bridges, towers, flyovers (overpasses), tunnels, off shore structures like oil and gas fields in the sea, aerostructure and other structures. This involves identifying the loads which act upon a structure and the forces and stresses which arise within that structure due to those loads, and then designing the structure to successfully support and resist those loads. The loads can be self weight of the structures, other dead load, live loads, moving (wheel) load, wind load, earthquake load, load from temperature change etc. The structural engineer must design structures to be safe for their users and to successfully fulfill the function they are designed for (to be *serviceable*). Due to the nature of some loading conditions, sub-disciplines within structural engineering have emerged, including wind engineering and earthquake engineering.

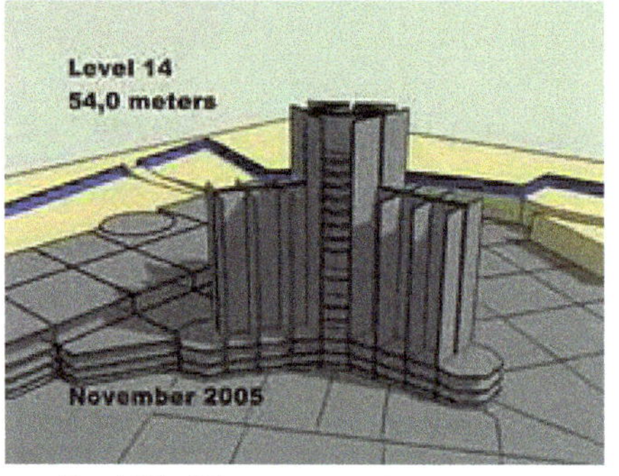

Burj Khalifa construction process

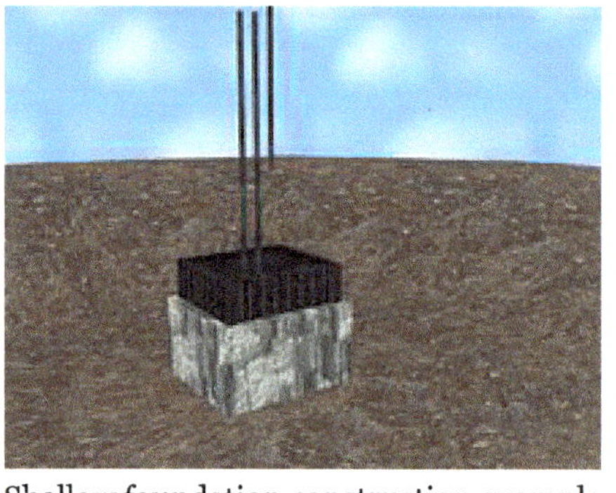

Shallow foundation construction example

Design considerations will include strength, stiffness, and stability of the structure when subjected to loads which may be static, such as furniture or self-weight, or dynamic, such as wind, seismic, crowd or vehicle loads, or transitory, such as temporary construction loads or impact. Other considerations include cost, constructability, safety, aesthetics and sustainability.

Surveying

A student using a theodolite

A student using a dumpy level

Surveying is the process by which a surveyor measures certain dimensions that occur on or near the surface of the Earth. Surveying equipment, such as levels and theodolites, are used for accurate measurement of angular deviation, horizontal, vertical and slope distances. With computerisation, electronic distance measurement (EDM), total stations, GPS surveying and laser scanning have to a large extent supplanted traditional instruments. Data collected by survey measurement is converted into a graphical representation of the Earth's surface in the form of a map. This information is then used by civil engineers, contractors and realtors to design from, build on, and trade, respectively. Elements of a structure must be sized and positioned in relation to each other and to site boundaries and adjacent structures. Although surveying is a distinct profession with separate qualifications and licensing arrangements, civil engineers are trained in the basics of surveying and mapping, as well as geographic information systems. Surveyors also lay out the routes of railways, tramway tracks, highways, roads, pipelines and streets as well as position other infrastructure, such as harbors, before construction.

Land Surveying

BLM cadastral survey marker from 1992 in San Xavier, Arizona.

In the United States, Canada, the United Kingdom and most Commonwealth countries land surveying is considered to be a separate and distinct profession. Land surveyors are not considered to be engineers, and have their own professional associations and licensing requirements. The services of a licensed land surveyor are generally required for boundary surveys (to establish the boundaries of a parcel using its legal description) and subdivision plans (a plot or map based on a survey of a parcel of land, with boundary lines drawn inside the larger parcel to indicate the creation of new boundary lines and roads), both of which are generally referred to as Cadastral surveying.

Construction Surveying

Construction surveying is generally performed by specialised technicians. Unlike land surveyors, the resulting plan does not have legal status. Construction surveyors perform the following tasks:

- Surveying existing conditions of the future work site, including topography, existing buildings and infrastructure, and underground infrastructure when possible;
- "lay-out" or "setting-out": placing reference points and markers that will guide the construction of new structures such as roads or buildings;
- Verifying the location of structures during construction;
- As-Built surveying: a survey conducted at the end of the construction project to verify that the work authorized was completed to the specifications set on plans.

Transportation Engineering

The engineering of this roundabout in Bristol, England, attempts to make traffic flow free-moving

Transportation engineering is concerned with moving people and goods efficiently, safely, and in a manner conducive to a vibrant community. This involves specifying, designing, constructing, and maintaining transportation infrastructure which includes streets, canals, highways, rail systems, airports, ports, and mass transit. It includes areas such as transportation design, transportation planning, traffic engineering, some aspects of urban engineering, queueing theory, pavement engineering, Intelligent Transportation System (ITS), and infrastructure management.

Forensic Engineering

Forensic engineering is the investigation of materials, products, structures or components that fail or do not operate or function as intended, causing personal injury or damage to property. The consequences of failure are dealt with by the law of product liability. The field also deals with retracing processes and procedures leading to accidents in operation of vehicles or machinery. The subject is applied most commonly in civil law cases, although it may be of use in criminal law cases. Generally the purpose of a Forensic engineering investigation is to locate cause or causes of failure with a view to improve performance or life of a component, or to assist a court in determining the facts of an accident. It can also involve investigation of intellectual property claims, especially patents.

Municipal or Urban Engineering

Lake Chapultepec

Municipal engineering is concerned with municipal infrastructure. This involves specifying, designing, constructing, and maintaining streets, sidewalks, water supply networks, sewers, street lighting, municipal solid waste management and disposal, storage depots for various bulk materials used for maintenance and public works (salt, sand, etc.), public parks and cycling infrastructure. In the case of underground utility networks, it may also include the civil portion (conduits and access chambers) of the local distribution networks of electrical and telecommunications services. It can also include the optimizing of waste collection and bus service networks. Some of these disciplines overlap with other civil engineering specialties, however municipal engineering focuses on the coordination of these infrastructure networks and services, as they are often built simultaneously, and managed by the same municipal authority. Municipal engineers may also design the site civil works for large buildings, industrial plants or campuses (i.e. access roads, parking lots, potable water supply, treatment or pretreatment of waste water, site drainage, etc.)

Control Engineering

Control engineering (or *control systems engineering*) is the branch of civil engineering discipline that applies control theory to design systems with desired behaviors. The practice uses sensors to measure the output performance of the device being controlled (often a vehicle) and those measurements can be used to give feedback to the input actuators that can make corrections toward desired performance. When a device is designed to perform without the need of human inputs for correction it is called automatic control (such as cruise control for regulating a car's speed). Multidisciplinary in nature, control systems engineering activities focus on implementation of control systems mainly derived by mathematical modeling of systems of a diverse range.

Applications in Civil Engineering

Broadly, civil Engineering curriculum covers structural, geotechnical, transportation, water resources, and environmental engineering related subjects. In most of these subjects, it is necessary to deal with the physical, mathematical, and simulation models and representative field data. When dealing with the models one has to deal with the experimental data. In many of these branches field data are collected to analyze and model the real life phenomenon and there are lot of uncertainties involved related to this data. Some of them are briefly described in the following paragraphs.

Concrete is the major building material and various grades of concrete are available for various applications. Compressive strength of a particular grade of concrete is determined based on the strength of a few sample cubes made from it. It is necessary to test a minimum number of cubes to have certain confidence on the characteristic strength of a particular grade of concrete. In this case, the test on a particular cube made from a particular grade of concrete is an experiment and when similar experiments are conducted on many samples the resulting strength may take different values. Though a particular gradation of materials is used in making these cubes, there is going to be some variability in the gradation of materials and there is no control on this variability. When huge quantity of concrete is mixed at once, it is difficult to control the variability in the gradation. This variability leads to different values of compressive strength. Other graphical representations of the same data are shown in the figures.

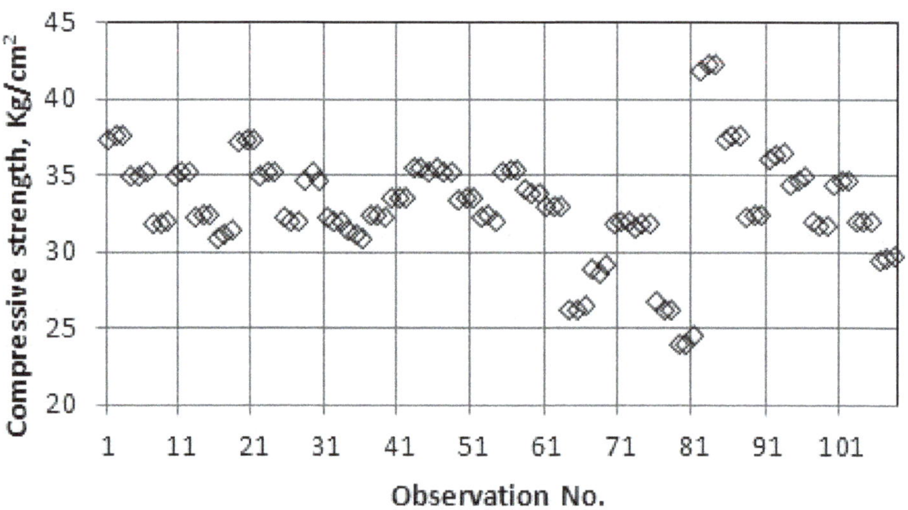

Compressive strength data of a particular grade of concrete

The mean compressive of strength of a representative sample cubes is taken as the characteristic strength of a particular grade of concrete. Strength of not more than 5% of cubes is expected to fall below the characteristic strength. The strength of the cubes made from a particular grade of concrete is a variable and hence the mean strength of a sample cubes is also a variable under repeated sampling. To understand this variability and its implications it is necessary to have minimum knowledge of probability methods and statistics.

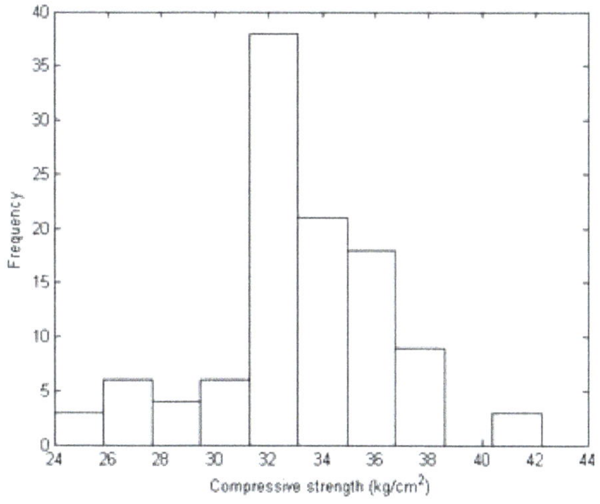

Frequency histogram of the compressive strength data

For example, if we know that the flexural strength of a concrete mix is normally distributed with σ = 1.5 MPa, and we want to estimate a Confidence Interval for the true mean (μ, mean flexural strength of that particular concrete mix) with 95% confidence and interval width of 1.0 MPa what are the minimum concrete beams to be tested? (n=?). Another example is to compare the effect of two admixtures. Is admixture A's mean effect on concrete strength equivalent to Admixture B's effect ?

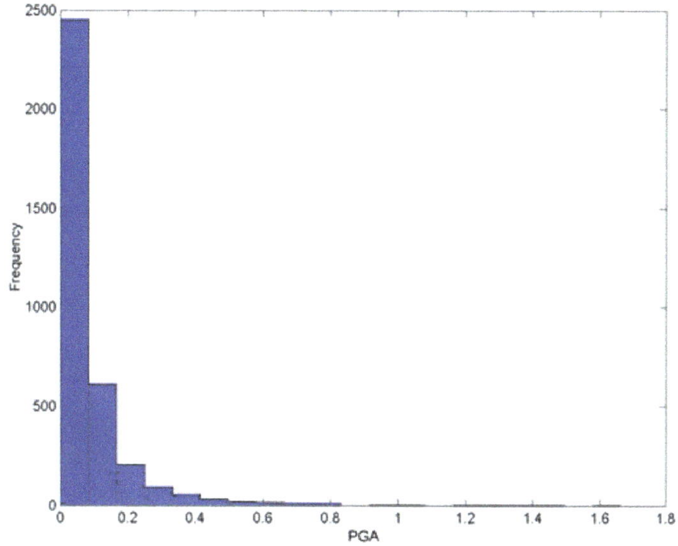

Frequency histogram of Peak Ground Acceleration data

Similarly, there are other areas of structural engineering where the probability concepts are equally important. Forces like wind force, earth quake force acting on the structures are variable and this variability also cannot be explained without out the support of relevant data. It is necessary to have certain understanding on this variability when using these forces in designing the structures.

For example, peak ground accelerations (PGA) are modeled using clipped normal, exponential distributions. To fit some kind of model it is necessary to understand the variability in the data.

Similarly, Geotechnical engineer has to deal with the natural materials and understanding of the properties of these materials is not comprehensive. These properties must be inferred from limited and costly observations. Offshore structures are the typical examples where many uncertainties, such as uncertainties about loads acting, and how much load a foundation can sustain. Probability concepts are important when designing the foundations on weak soils and when the soil conditions are varying. Various types of tests are required to evaluate the soil conditions. Minimum number of tests required in order to have certain confidence on the evaluated soil parameters is a question here. It is the responsibility of the engineer to interpret the limited amount of data that was collected. Probability concepts are useful in finding the answers for this kind of problems. Similarly, in slope stabilization, different kinds of additive materials are used and it is necessary to carry out comparative studies before selecting one.

In water resource engineering, in most of the sub-areas, basic understanding of probability concepts is necessary. In this field the uncertainties involved can be divided into four basic categories: hydrologic, hydraulic, structural, and economic. More specifically, uncertainties could arise from various sources including natural uncertainties, model uncertainties, parameter uncertainties, data uncertainties, and operational uncertainties. Despite numerous research efforts made to further the understanding of various processes in hydro systems, there is still much more that are beyond the researcher's grasp. Therefore, uncertainties exist due to lack of perfect knowledge concerning the phenomena and processes involved in the problem solving approach.

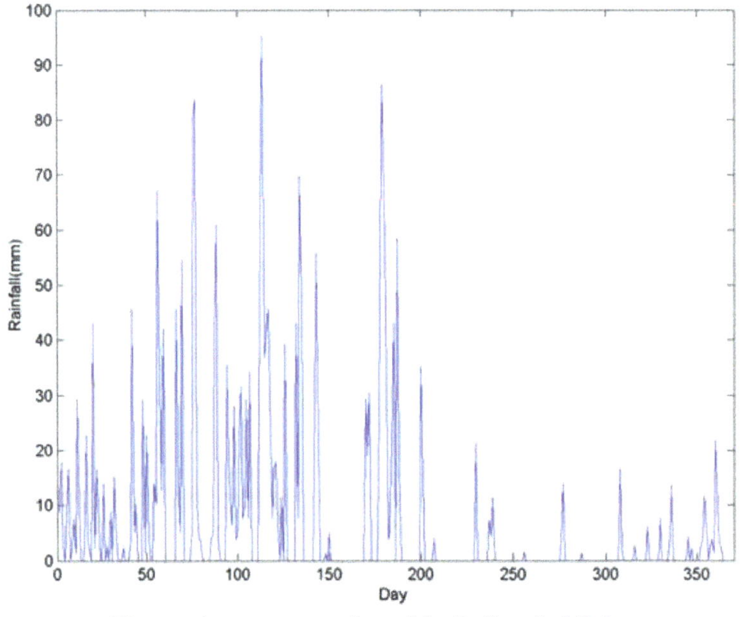

Time series representation of the Daily rainfall data

When constructing dams, cross drainage structures, it is necessary to predict the water levels for which these structures are to be constructed. Predicting the design forces that are going to act on these structures is a complex problem. Rainfall forecasting is one area where most of the things are uncertain and without having sound knowledge in the probability concepts this process cannot be carried out. The following figure shows the daily rainfall data over a one year period and from this figure it can be said that there is a lot of variability in the data.

For example, a dam is to be designed to safely retain the water in a reservoir. As shown in Figure, let H be the maximum water level in the reservoir in a generic year, and F is the associated horizontal force acting on a 1 meter length of the dam.

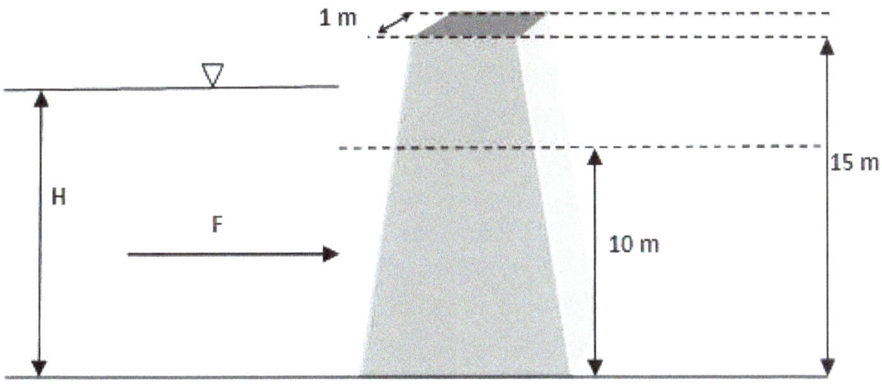

Schematic diagram of the forces acting on the dam

Assuming that F and H are related to each other and by taking a suitable probability distribution for H, it is necessary to evaluate the design value of the horizontal force acting on the dam. Similarly, the runoff data of river have lot of variability and the understanding of this variability is utmost important in designing the cross drainage structures.

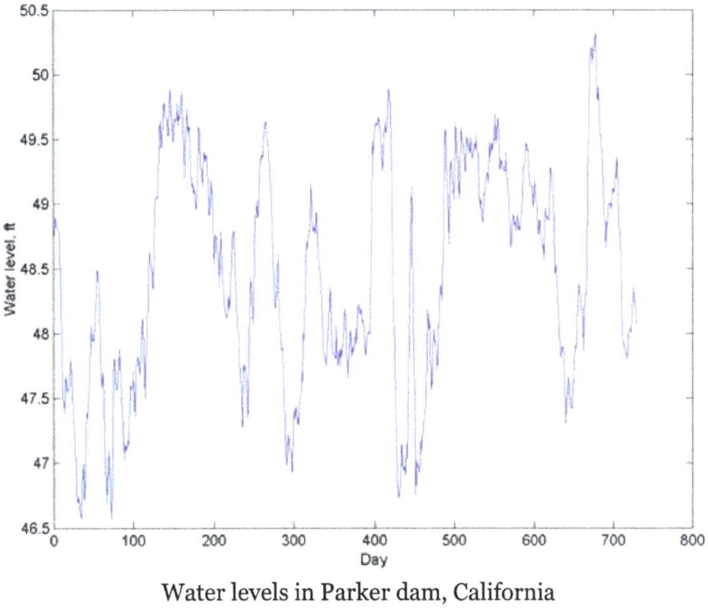

Water levels in Parker dam, California

Same is the problem when dealing with the problems in environmental engineering. To test the amount of bacteria, E-coli, metals etc, present in the water require thorough testing. These quantities may vary across geometries and this variation depends on several unknown factors. It is necessary to fix the number of tests that are to be conducted to evaluate the water quality.

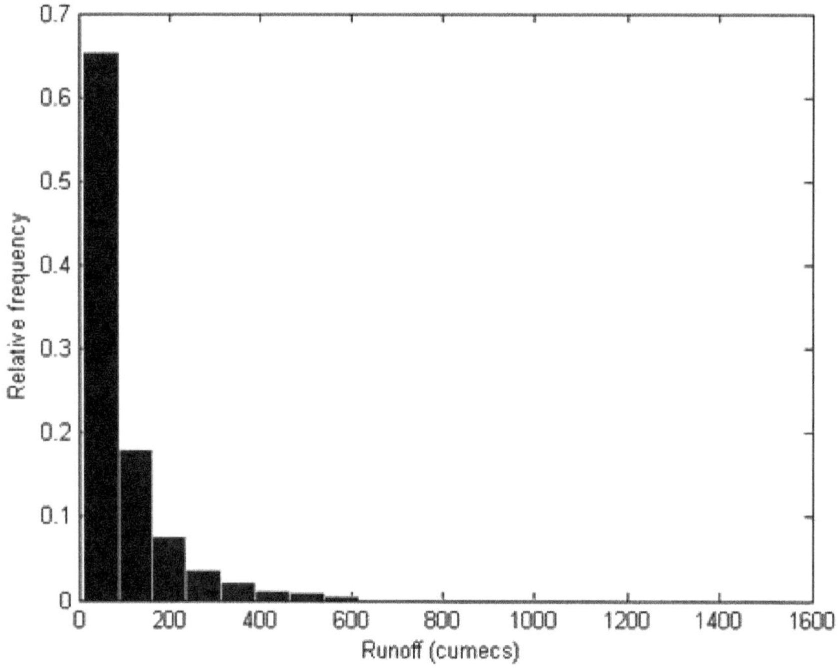

Relative frequency histogram of the runoff data observed in a river

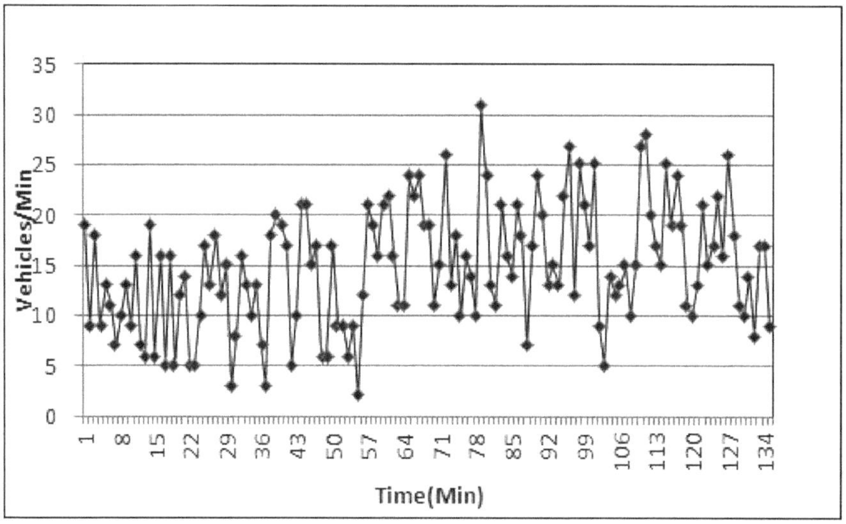

Time series data of vehicular flow observed at a road section

Human element causes lot of uncertainty when dealing with the transportation engineering related problems. Specifically this is significant when dealing with traffic engineering and transportation planning related problems. As shown in the figure, there is a huge variability in the number of vehicles crossing a road section in unit time period. Given the same conditions the same driver with the same vehicle may maintain different speeds at two different time instances. Same is true with the gap maintaining behavior with respect to the front vehicles, and the overtaking behavior, to quote a few situations. Knowledge about the vehicle arrivals plays important role in formulating the traffic management schemes. But this is a variable process and one may have to settle for highly probabilistic process and the design of the traffic control scheme has to be carried out on this basis.

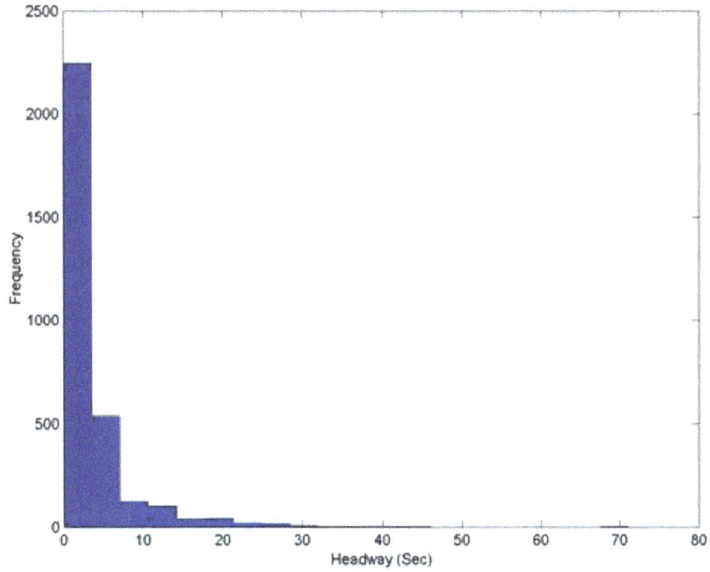

Headway frequencies observed on a single lane road

Under similar circumstances a person may choose different modes for his intercity movement. Normally this is attributed to the inherent randomness in the human being but there is an element of chance that this behavior might have resulted from our inability to understand the factors behind the mode choice. Knowledge of statistics is very important when dealing with transportation planning where policy makers/researchers use sample data in formulating the policies meant for future. Collecting the sample data is one of costliest processes involved in the transportation planning process. Mistakes committed at this stage, due to lack of statistics knowledge, renders the entire planning process invalid.

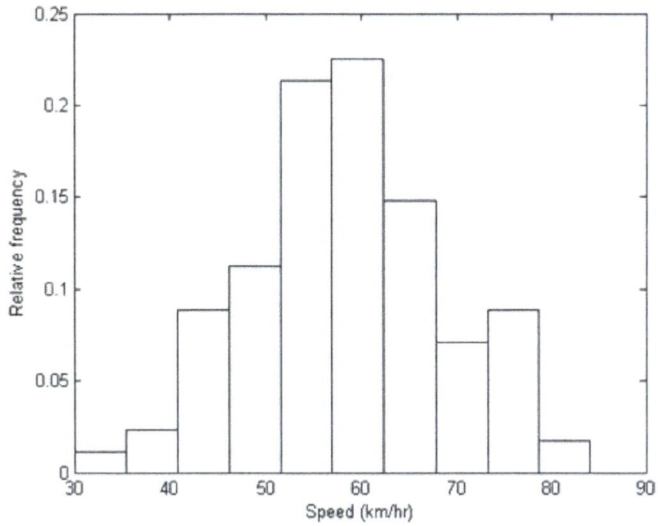

Relative frequency histogram of the free flow speed data of a particular vehicle type

Scope and Objectives of this Course

To begin with, the basic concepts of probability and statistics are discussed. Keeping in view the problems encountered in various branches of civil engineering, application oriented examples are provided along with the basic concepts. To make it more relevant to the civil engineering students of both under graduate and post graduate level, examples from all major civil engineering branches are discussed. Wherever possible, experimental data have been utilized in the example problems.

Event (Probability Theory)

In probability theory, an event is a set of outcomes of an experiment (a subset of the sample space) to which a probability is assigned. A single outcome may be an element of many different events, and different events in an experiment are usually not equally likely, since they may include very different groups of outcomes. An event defines a complementary event, namely the complementary set (the event *not* occurring), and together these define a Bernoulli trial: did the event occur or not?

Typically, when the sample space is finite, any subset of the sample space is an event (*i.e.* all elements of the power set of the sample space are defined as events). However, this approach does not work well in cases where the sample space is uncountably infinite, most notably when the outcome is a real number. So, when defining a probability space it is possible, and often necessary, to exclude certain subsets of the sample space from being events.

A Simple Example

If we assemble a deck of 52 playing cards with no jokers, and draw a single card from the deck, then the sample space is a 52-element set, as each card is a possible outcome. An event, however, is any subset of the sample space, including any singleton set (an elementary event), the empty set (an impossible event, with probability zero) and the sample space itself (a certain event, with probability one). Other events are proper subsets of the sample space that contain multiple elements. So, for example, potential events include:

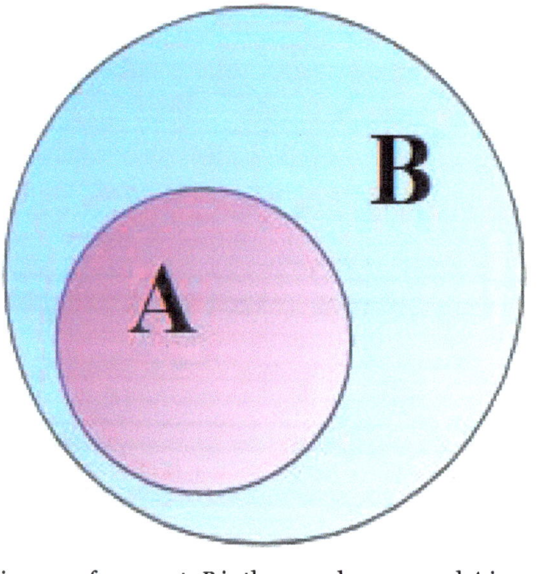

A Euler diagram of an event. *B* is the sample space and *A* is an event. By the ratio of their areas, the probability of *A* is approximately 0.4.

- "Red and black at the same time without being a joker" (0 elements),
- "The 5 of Hearts" (1 element),
- "A King" (4 elements),
- "A Face card" (12 elements),
- "A Spade" (13 elements),
- "A Face card or a red suit" (32 elements),
- "A card" (52 elements).

Since all events are sets, they are usually written as sets (e.g. {1, 2, 3}), and represented graphically using Venn diagrams. In the situation where each outcome in the sample space Ω is equally likely, the probability P of an event A is the following formula:

$$\mathrm{P}(A) = \frac{|A|}{|\Omega|} \left(\text{alternatively: } \Pr(A) = \frac{|A|}{|\Omega|} \right)$$

This rule can readily be applied to each of the example events above.

Events in Probability Spaces

Defining all subsets of the sample space as events works well when there are only finitely many outcomes, but gives rise to problems when the sample space is infinite. For many standard probability distributions, such as the normal distribution, the sample space is the set of real numbers or some subset of the real numbers. Attempts to define probabilities for all subsets of the real numbers run into difficulties when one considers 'badly behaved' sets, such as those that are nonmeasurable. Hence, it is necessary to restrict attention to a more limited family of subsets. For the standard tools of probability theory, such as joint and conditional probabilities, to work, it is necessary to use a σ-algebra, that is, a family closed under complementation and countable unions of its members. The most natural choice is the Borel measurable set derived from unions and intersections of intervals. However, the larger class of Lebesgue measurable sets proves more useful in practice.

In the general measure-theoretic description of probability spaces, an event may be defined as an element of a selected σ-algebra of subsets of the sample space. Under this definition, any subset of the sample space that is not an element of the σ-algebra is not an event, and does not have a probability. With a reasonable specification of the probability space, however, all *events of interest* are elements of the σ-algebra.

A Note on Notation

Even though events are subsets of some sample space Ω, they are often written as propositional formulas involving random variables. For example, if X is a real-valued random variable defined on the sample space Ω, the event

$$\{\omega \in \Omega \mid u < X(\omega) \leq v\}$$

can be written more conveniently as, simply,

$$u < X \le v.$$

This is especially common in formulas for a probability, such as

$$\Pr(u < X \le v) = F(v) - F(u).$$

The set $u < X \le v$ is an example of an inverse image under the mapping X because $\omega \in X^{-1}((u, v])$ if and only if $u < X(\omega) \le v$.

Random Events

Experimental or the field data can be described using simple descriptors which are useful in understanding the uncertainty and the variability in the data. These descriptors alone may not be useful in utilizing this data in real life applications such as describing some natural phenomenon. For effective utilization of the experimental data, knowledge in statistics is necessary. Statistics itself is based on the concepts of probability and it is necessary to have sound understanding in the fundamentals of probability theory. In this chapter, the basic data descriptors, such as graphical and numerical descriptors, are briefly discussed before going to the basic concepts of probability. Set theory applications in the probability are discussed briefly followed by the classical probability axioms.

Data Description

Data can be represented in various forms and this representation enables better understanding of the process or phenomenon on which the data has been collected. Graphical and numerical descriptions are the two forms in which the data can be represented.

Graphical Representation

When the data is represented graphically it is easier to see and understand the variability in the data set, which otherwise is difficult to see. Histograms and frequency diagrams are some of the important graphical forms using which the data can be represented. Understanding the variability in the data through graphical means is the basis for the application of probability concepts.

Histogram

A histogram is a graphical representation of the distribution of numerical data. It is an estimate of the probability distribution of a continuous variable (quantitative variable) and was first introduced by Karl Pearson. It is a kind of bar graph. To construct a histogram, the first step is to "bin" the range of values—that is, divide the entire range of values into a series of intervals—and then count how many values fall into each interval. The bins are usually specified as consecutive, non-overlapping intervals of a variable. The bins (intervals) must be adjacent, and are often (but are not required to be) of equal size.

If the bins are of equal size, a rectangle is erected over the bin with height proportional to the frequency — the number of cases in each bin. A histogram may also be normalized to display "relative" frequencies. It then shows the proportion of cases that fall into each of several categories, with the sum of the heights equaling 1.

However, bins need not be of equal width; in that case, the erected rectangle is defined to have its *area* proportional to the frequency of cases in the bin. The vertical axis is then not the frequency but *frequency density* — the number of cases per unit of the variable on the horizontal axis. Examples of variable bin width are displayed on Census bureau data below.

As the adjacent bins leave no gaps, the rectangles of a histogram touch each other to indicate that the original variable is continuous.

Histograms give a rough sense of the density of the underlying distribution of the data, and often for density estimation: estimating the probability density function of the underlying variable. The total area of a histogram used for probability density is always normalized to 1. If the length of the intervals on the x-axis are all 1, then a histogram is identical to a relative frequency plot.

A histogram can be thought of as a simplistic kernel density estimation, which uses a kernel to smooth frequencies over the bins. This yields a smoother probability density function, which will in general more accurately reflect distribution of the underlying variable. The density estimate could be plotted as an alternative to the histogram, and is usually drawn as a curve rather than a set of boxes.

Another alternative is the average shifted histogram, which is fast to compute and gives a smooth curve estimate of the density without using kernels.

The histogram is one of the seven basic tools of quality control.

Histograms are sometimes confused with bar charts. A histogram is used for continuous data, where the bins represent ranges of data, while a bar chart is a plot of categorical variables. Some authors recommend that bar charts have gaps between the rectangles to clarify the distinction.

Etymology

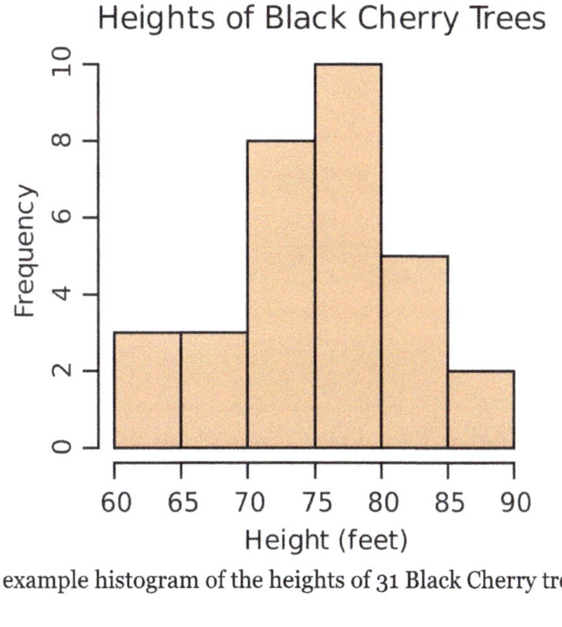

An example histogram of the heights of 31 Black Cherry trees.

The etymology of the word *histogram* is uncertain. Sometimes it is said to be derived from the Ancient Greek – "anything set upright" (as the masts of a ship, the bar of a loom, or the vertical bars of a histogram); and (*gramma*) – "drawing, record, writing". It is also said that Karl Pearson, who introduced the term in 1891, derived the name from "historical diagram".

Examples

This is a toy example:

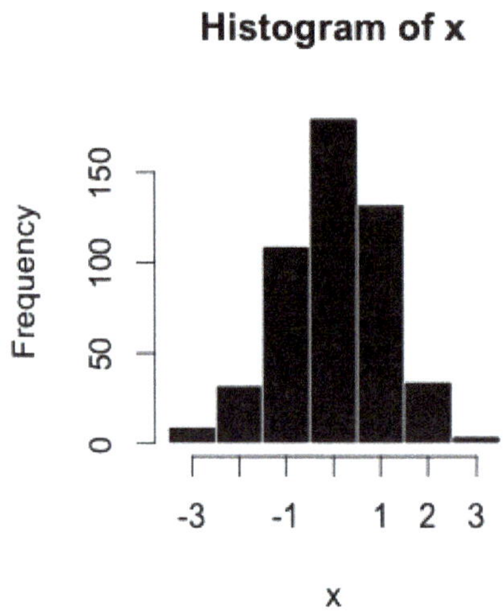

Bin	Count
−3.5	23
−2.5	32
−1.5	109
−0.5	180
0.5	132
1.5	34
2.5	4
3.5	90

The words used to describe the patterns in a histogram are: "symmetric", "skewed left" or "right", "unimodal", "bimodal" or "multimodal".

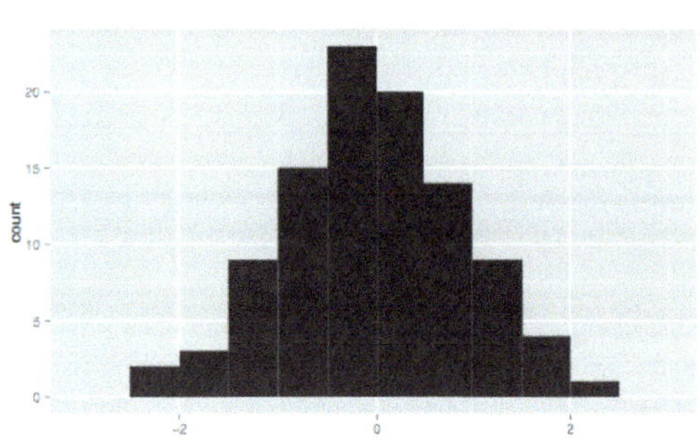

Symmetric, unimodal

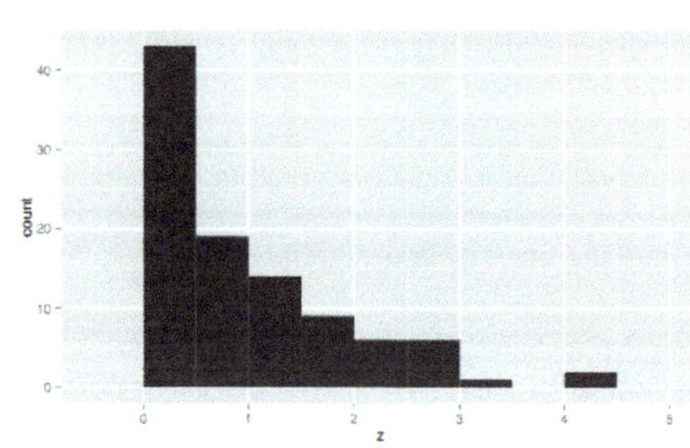

Skewed right

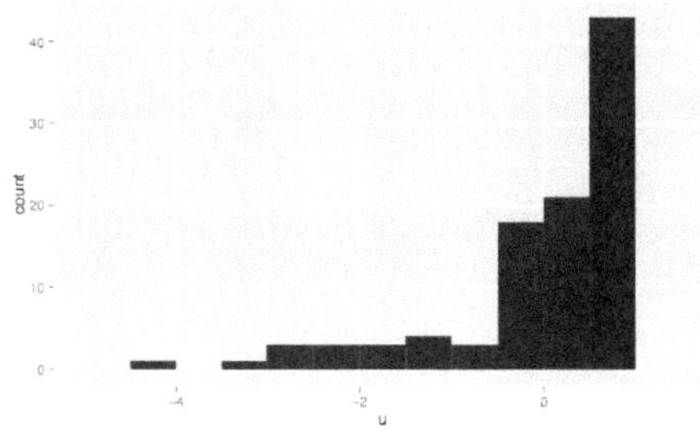

Skewed left

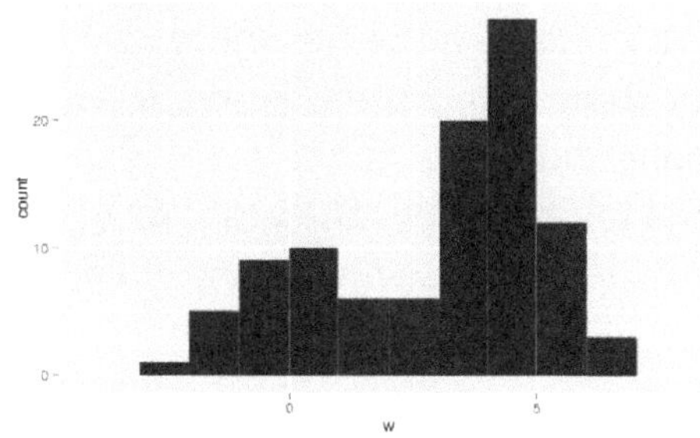

Bimodal

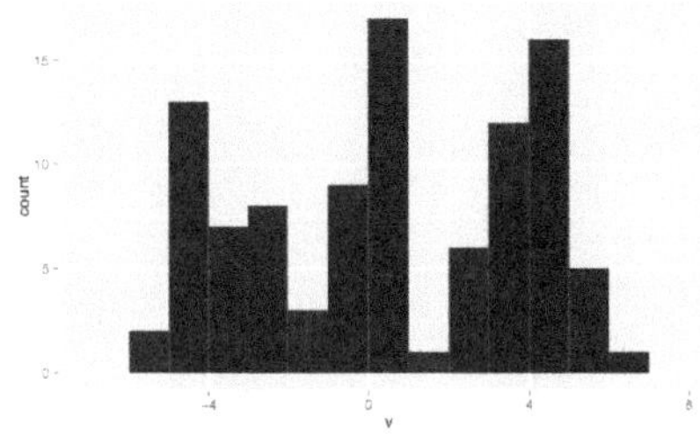

Multimodal

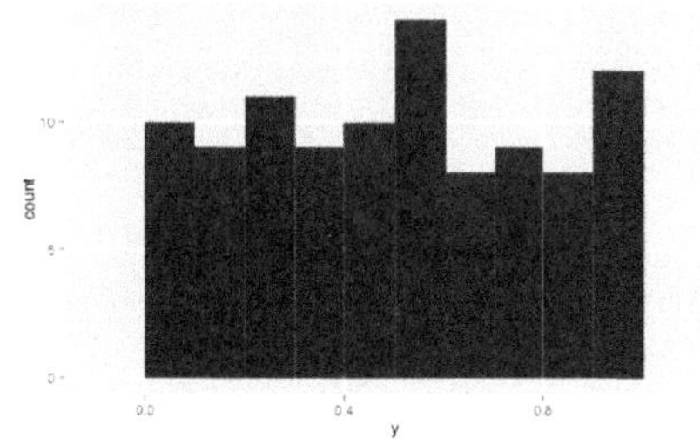

Symmetric

It is a good idea to plot the data using several different bin widths to learn more about it. Here is an example on tips given in a restaurant.

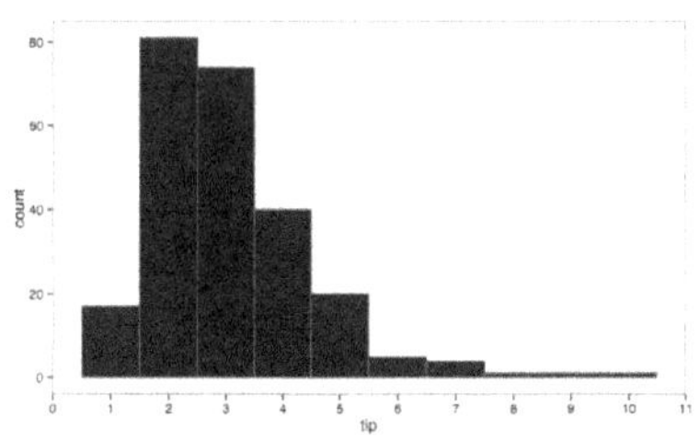

Tips using a $1 bin width, skewed right, unimodal

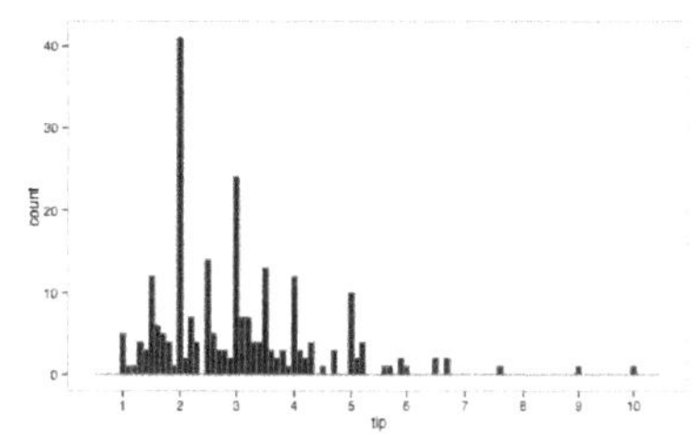

Tips using a 10c bin width, still skewed right, multimodal with modes at $ and 50c amounts, indicates rounding, also some outliers

Here are a couple more examples:

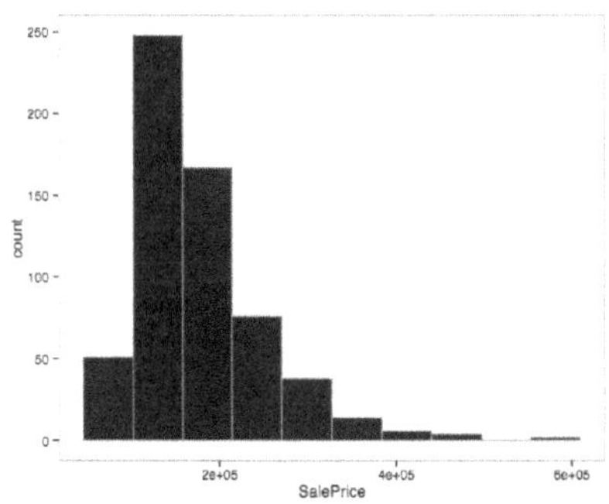

Prices of houses sold in Ames in 2009 exhibits some right skew

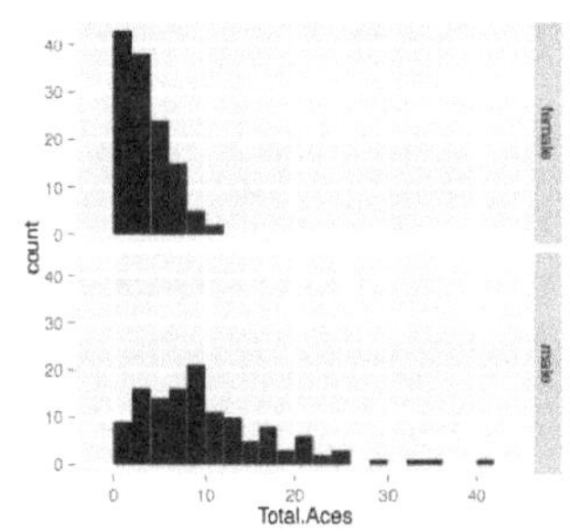

Aces by players in a grand slam tennis tournament, facetted by gender. There are more aces in the men's game.

The U.S. Census Bureau found that there were 124 million people who work outside of their homes. Using their data on the time occupied by travel to work, the table below shows the absolute number of people who responded with travel times "at least 30 but less than 35 minutes" is higher than the numbers for the categories above and below it. This is likely due to people rounding their reported journey time. The problem of reporting values as somewhat arbitrarily rounded numbers is a common phenomenon when collecting data from people.

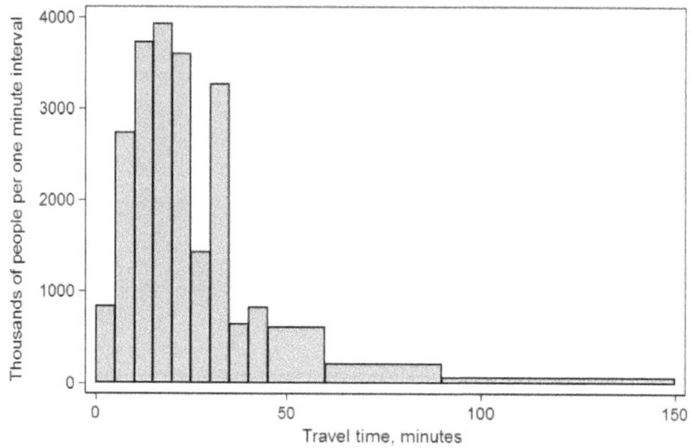

Histogram of travel time (to work), US 2000 census. Area under the curve equals the total number of cases. This diagram uses Q/width from the table.

Data by absolute numbers			
Interval	**Width**	**Quantity**	**Quantity/width**
0	5	4180	836
5	5	13687	2737
10	5	18618	3723
15	5	19634	3926
20	5	17981	3596
25	5	7190	1438
30	5	16369	3273
35	5	3212	642
40	5	4122	824
45	15	9200	613
60	30	6461	215
90	60	3435	57

This histogram shows the number of cases per unit interval as the height of each block, so that the area of each block is equal to the number of people in the survey who fall into its category. The area under the curve represents the total number of cases (124 million). This type of histogram shows absolute numbers, with Q in thousands.

Data by proportion			
Interval	**Width**	**Quantity (Q)**	**Q/total/width**
0	5	4180	0.0067
5	5	13687	0.0221

10	5	18618	0.0300
15	5	19634	0.0316
20	5	17981	0.0290
25	5	7190	0.0116
30	5	16369	0.0264
35	5	3212	0.0052
40	5	4122	0.0066
45	15	9200	0.0049
60	30	6461	0.0017
90	60	3435	0.0005

This histogram differs from the first only in the vertical scale. The area of each block is the fraction of the total that each category represents, and the total area of all the bars is equal to 1 (the fraction meaning "all"). The curve displayed is a simple density estimate. This version shows proportions, and is also known as a unit area histogram.

In other words, a histogram represents a frequency distribution by means of rectangles whose widths represent class intervals and whose areas are proportional to the corresponding frequencies: the height of each is the average frequency density for the interval. The intervals are placed together in order to show that the data represented by the histogram, while exclusive, is also contiguous. (E.g., in a histogram it is possible to have two connecting intervals of 10.5–20.5 and 20.5–33.5, but not two connecting intervals of 10.5–20.5 and 22.5–32.5. Empty intervals are represented as empty and not skipped.)

Mathematical Definition

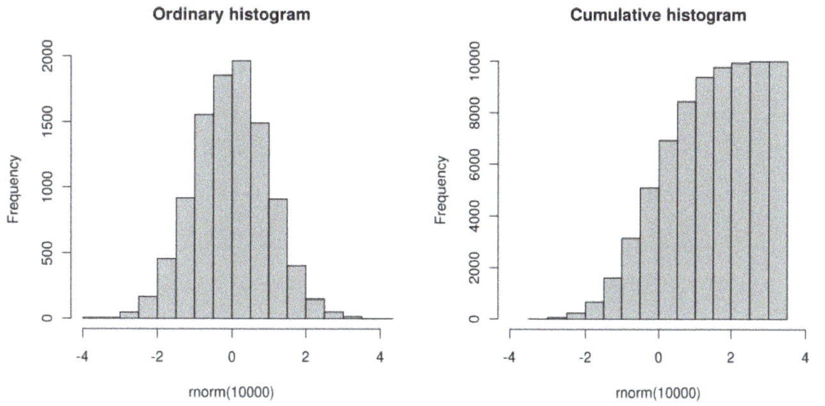

An ordinary and a cumulative histogram of the same data. The data shown is a random sample of 10,000 points from a normal distribution with a mean of 0 and a standard deviation of 1.

In a more general mathematical sense, a histogram is a function m_i that counts the number of observations that fall into each of the disjoint categories (known as *bins*), whereas the graph of a histogram is merely one way to represent a histogram. Thus, if we let n be the total number of observations and k be the total number of bins, the histogram m_i meets the following conditions:

$$n = \sum_{i=1}^{k} m_i.$$

Cumulative Histogram

A cumulative histogram is a mapping that counts the cumulative number of observations in all of the bins up to the specified bin. That is, the cumulative histogram M_i of a histogram m_j is defined as:

$$M_i = \sum_{j=1}^{i} m_j.$$

Number of Bins and Width

There is no "best" number of bins, and different bin sizes can reveal different features of the data. Grouping data is at least as old as Graunt's work in the 17th century, but no systematic guidelines were given until Sturges's work in 1926.

Using wider bins where the density is low reduces noise due to sampling randomness; using narrower bins where the density is high (so the signal drowns the noise) gives greater precision to the density estimation. Thus varying the bin-width within a histogram can be beneficial. Nonetheless, equal-width bins are widely used.

Some theoreticians have attempted to determine an optimal number of bins, but these methods generally make strong assumptions about the shape of the distribution. Depending on the actual data distribution and the goals of the analysis, different bin widths may be appropriate, so experimentation is usually needed to determine an appropriate width. There are, however, various useful guidelines and rules of thumb.

The number of bins k can be assigned directly or can be calculated from a suggested bin width h as:

$$k = \left\lceil \frac{\max x - \min x}{h} \right\rceil.$$

The braces indicate the ceiling function.

Square-root Choice

$$k = \sqrt{n},$$

which takes the square root of the number of data points in the sample (used by Excel histograms and many others).

Sturges' Formula

Sturges' formula is derived from a binomial distribution and implicitly assumes an approximately normal distribution.

$$k = \langle \log_2 n \rangle + 1,$$

It implicitly bases the bin sizes on the range of the data and can perform poorly if $n < 30$, because the number of bins will be small—less than seven—and unlikely to show trends in the data well. It may also perform poorly if the data are not normally distributed.

Rice Rule

$$k = \langle 2n^{1/3} \rangle,$$

The Rice Rule is presented as a simple alternative to Sturges's rule.

Doane's Formula

Doane's formula is a modification of Sturges' formula which attempts to improve its performance with non-normal data.

$$k = 1 + \log_2(n) + \log_2\left(1 + \frac{|g_1|}{\sigma_{g_1}}\right)$$

where g_1 is the estimated 3rd-moment-skewness of the distribution and

$$\sigma_{g_1} = \sqrt{\frac{6(n-2)}{(n+1)(n+3)}}$$

Scott's Normal Reference Rule

$$h = \frac{3.5\hat{\sigma}}{n^{1/3}},$$

where $\hat{\sigma}$ is the sample standard deviation. Scott's normal reference rule is optimal for random samples of normally distributed data, in the sense that it minimizes the integrated mean squared error of the density estimate.

Freedman–Diaconis' Choice

The Freedman–Diaconis rule is:

$$h = 2\frac{\mathrm{IQR}(x)}{n^{1/3}},$$

which is based on the interquartile range, denoted by IQR. It replaces 3.5σ of Scott's rule with 2 IQR, which is less sensitive than the standard deviation to outliers in data.

Minimizing Cross-validation Estimated Squared Error

This approach of minimizing integrated mean squared error from Scott's rule can be generalized beyond Normal distributions, by using leave-one out cross validation:

$$\underset{h}{\operatorname{argmin}}\, \hat{J}(h) = \underset{h}{\operatorname{argmin}} \left(\frac{2}{(n-1)h} - \frac{n+1}{n^2(n-1)h} \sum_k N_k^2 \right)$$

Here, N_k is the number of datapoints in the kth bin, and choosing the value of h that minimizes J will minimize integrated mean squared error.

Choice Based on Minimization of an Estimated L2 Risk Function

$$\underset{h}{\operatorname{argmin}} \frac{2\bar{m} - v}{h^2}$$

where $\bar{m}$ and v are mean and biased variance of a histogram with bin-width h, $\bar{m} = \frac{1}{k}\sum_{i=1}^{k} m_i$ and $v = \frac{1}{k}\sum_{i=1}^{k} (m_i - \bar{m})^2$.

Remark

A good reason why the number of bins should be proportional to $n^{1/3}$ is the following: suppose that the data are obtained as n independent realizations of a bounded probability distribution with smooth density. Then the histogram remains equally »rugged« as n tends to infinity. If s is the »width« of the distribution (e. g., the standard deviation or the inter-quartile range), then the number of units in a bin (the frequency) is of order nh/s and the *relative* standard error is of order $\sqrt{s/(nh)}$. Comparing to the next bin, the relative change of the frequency is of order h/s provided that the derivative of the density is non-zero. These two are of the same order if h is of order $s/n^{1/3}$, so that k is of order $n^{1/3}$.

This simple cubic root choice can also be applied to bins with non-constant width.

When the data is limited (less than 25 data points), a line diagram can be used as a graphical form to represent the data. When data points are more than 25 it is useful to present it in the form of histogram.

The free flow speed data of the cars collected on the National Highway stretch in Guwahati, Assam, is shown in the table. Total number of observations is 169 and it has been observed that there is a considerable variation in the speeds and the maximum and minimum speeds are 84 and 30 km/hr, respectively. For displaying this data in a histogram, the data are divided into groups according to their magnitudes. Each group or class is of equal length and the length of the group is decided based on the range, the difference between the largest and smallest values over which the data are spread. The number of data points falling in each class or group is the frequency corresponding to that group. The number of classes should be at least 5 and not more than 25, and as a thumb rule it should be close to the square root of the number of data points. Data on car speeds is grouped into 11 classes and the corresponding frequency computations are shown in the table.

Table: Observed Speed data of Cars

61	64	64	48	49	64	77	40	68
77	61	73	52	53	65	42	55	68
55	56	57	50	62	65	51	61	51
58	60	64	63	54	43	57	69	80
63	62	59	62	55	67	44	55	57
74	61	54	50	57	47	53	68	68
67	56	75	64	74	65	74	40	76
76	60	50	55	50	52	78	64	55
62	62	61	59	84	64	51	62	66
67	61	77	34	60	42	45	58	
54	60	50	68	78	62	43	62	
66	51	61	76	30	49	58	65	
64	60	77	43	57	53	55	52	
50	62	50	51	43	67	40	53	
49	63	52	63	41	61	53	61	
59	60	58	62	44	42	44	55	
64	79	43	44	63	74	50	61	
74	53	55	53	72	54	59	55	
68	55	60	50	54	58	46	61	
68	70	55	69	56	60	37	65	

Absolute frequency versus the middle point of the class, in the form of the histogram, is shown in the figure. It is normally presumed that the free flow speed data follows normal probability distribution and the histogram shown in this figure more or less is indicating the same. This approach gives a basic idea on the accuracy of the assumption made. For the same data the relative frequency histogram is shown in the figure. In the following figure, the relative frequency polygon is shown. This polygon is obtained by joining the mid points of the blocks in the histogram. This is nothing but a probability curve when number of observed data points tends to infinity and the class widths are reduced, and this curve is known as probability density function of the population from which the data have been observed.

When the class width is not small the density function can be obtained by dividing the relative frequencies with the class width (known as the relative frequency density) and plotting these values in the form of a relative frequency polygon. Area of the relative frequency diagram shown in the following figure is not equals to one hence the relative frequencies are to be divided with the class width.

Table: Grouping of Speed data of cars and frequency computations

Class lower limit	Class upper limit	Class middle limit	Absolute frequency	Relative frequency	Absolute cumulative frequency
36	40	38	2	0.012	2
41	45	43	5	0.030	7
46	50	48	14	0.083	21
51	55	53	19	0.112	40
56	60	58	31	0.183	71
61	65	63	33	0.195	104
66	70	68	31	0.183	135
71	75	73	14	0.083	149
76	80	78	11	0.065	160
81	85	83	8	0.047	168
86	90	88	1	0.006	169

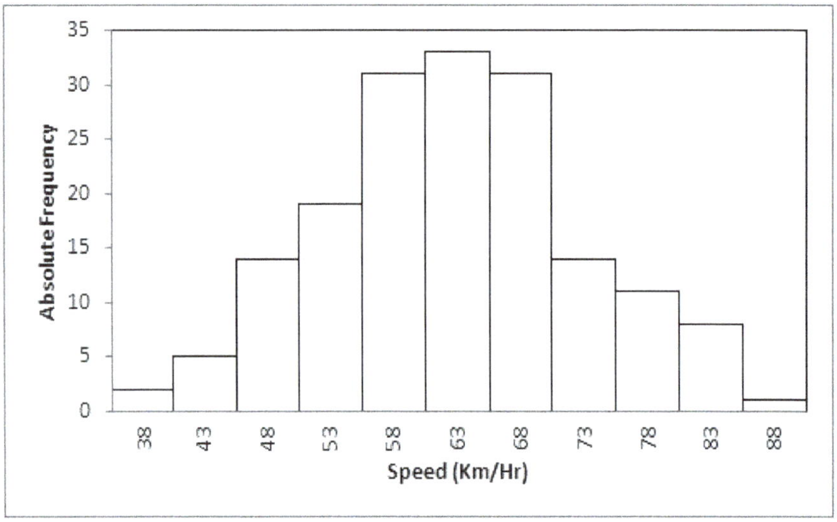

Frequency histogram of the free flow speed data of cars

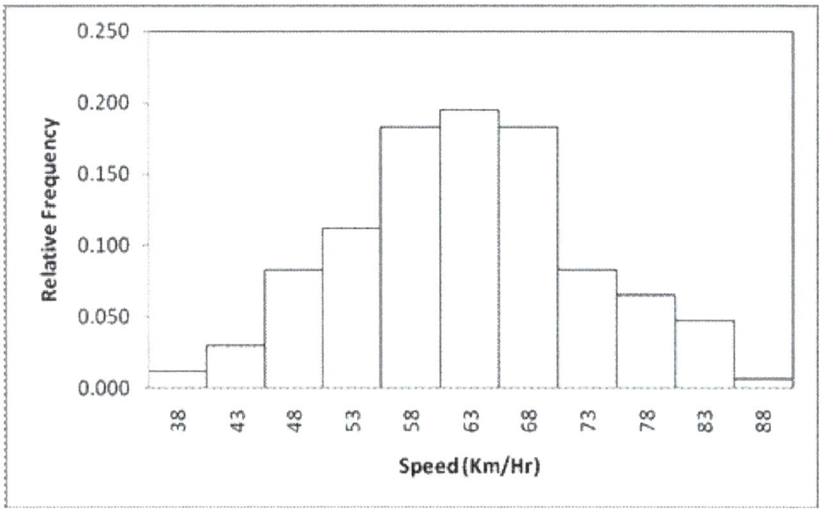

Relative frequency histogram of the free flow speed data of cars

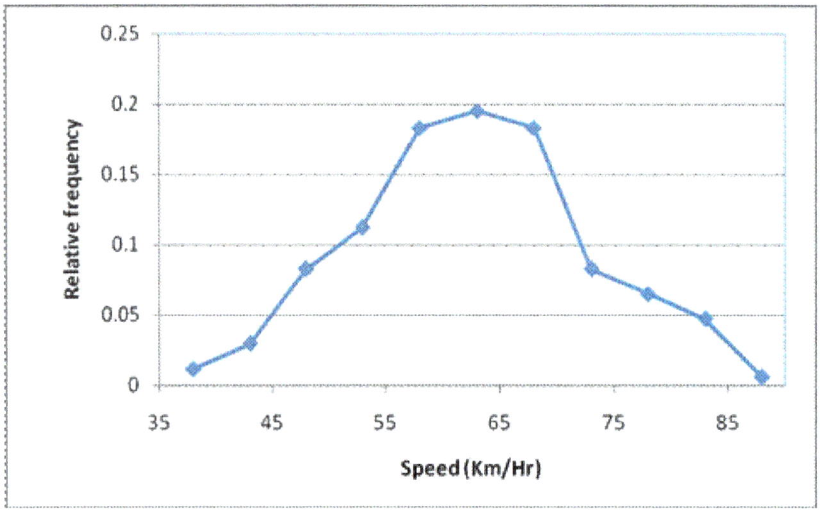

Relative frequency polygon of speed data of cars

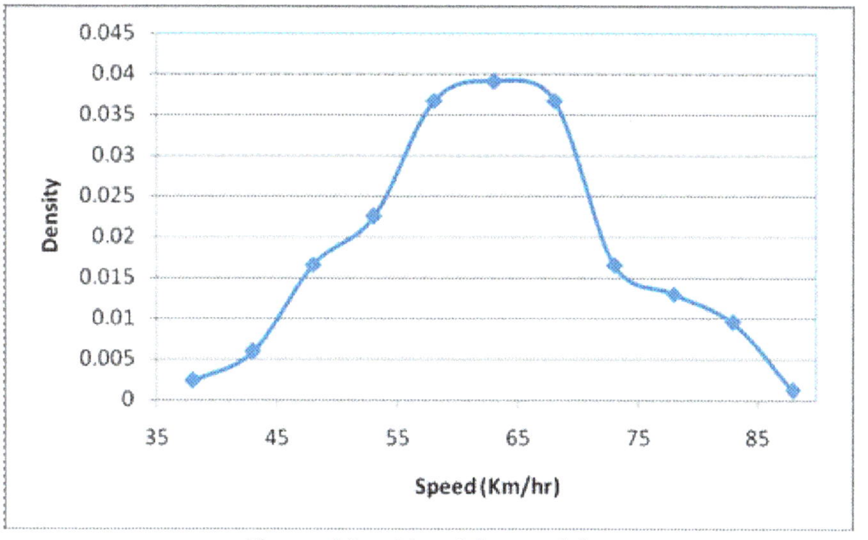

Observed densities of the speed data

Cumulative Frequency Diagrams

Cumulative frequency diagram is obtained by plotting the cumulative sum of the relative frequencies against the ends of the class boundaries. In the cumulative relative frequency diagrams, the ordinates represent the cumulative frequency of the occurrence of events less than or equal to the value on the abscissa. If the observed data points tend to infinity the resulting cumulative relative frequency diagram is called as cumulative distribution function. The following figure shows the cumulative relative frequency diagram of the speed data shown in the table. From the cumulative relative frequency diagrams it is easy to get the median and the other quartiles (median and the quartiles are covered in the numerical description of the data). A cumulative frequency diagram is also known as the *quantile* plot or the Q-plot. Quantiles are nothing but (n-1) values that divide the total frequency into n equal parts and in a quantile plot the qunatiles are shown on y-axis. Quantiles of two different data sets or the theoretical and the observed quantiles can be compared using the Q-Q plots.

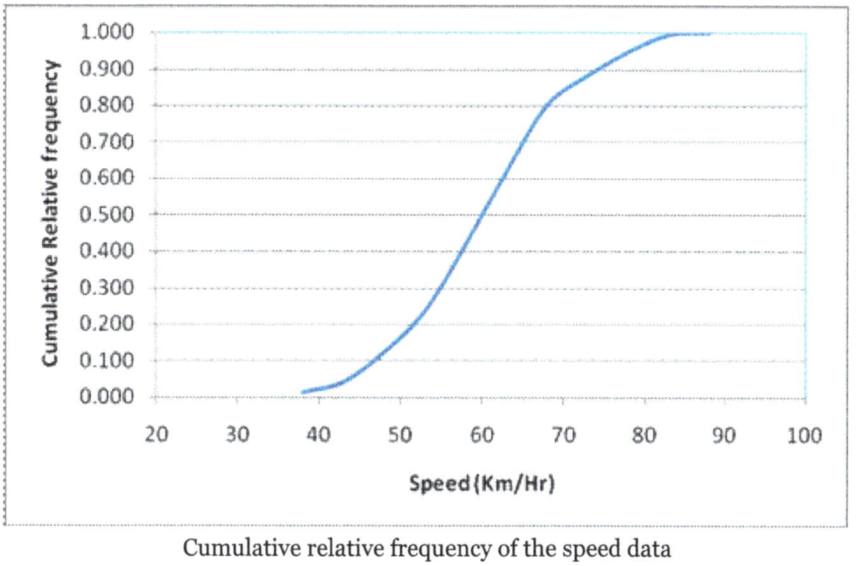

Cumulative relative frequency of the speed data

Numerical Description of the Data

Numerical description of the data is useful in describing the observed data in a more precise way and this method utilizes a set of characteristic numbers to describe the data. Numerical descriptors are useful in statistical inference that cannot be done using graphical measures. Three types of numerical descriptors, namely, measures of central tendency, dispersion, and asymmetry are commonly used to describe the data.

Measures of Central Tendency

Any observed data is spread around some numerical values and these values can be used to study the central tendency of the data. Mean, mode, and median are such numerical values that are useful in describing the central tendency of the data.

Mean

When a single value is used to describe the sample data, mean is ideal for this purpose. In general the notation $\bar{x}$ is used to denote sample mean and μ is used to denote the population mean. Simple arithmetic mean of a sample is calculated as follows;

$$\bar{x}=\frac{1}{n}\sum_{i=1}^{n}x_i$$

For the speed data presented in the table, the mean value is found to be 58.59 km/hr. For compressive strength data shown in the table, the mean strength is 33 MPa. For theoretical purposes mean is an important numerical measure of location. Mean is the centroid of the area under any frequency polygon when the sample size approaches infinity. Mean value calculated from any sample observations is prone to be influenced by too large or too small numerical values that are present in the data. These values are the deviations from the normal outcomes of the experimental results and might have resulted in due to the errors committed while conducting the experiment or in recording the experimental results. Under these conditions if the mean value is used to analyze the central tendency without looking into the data, it may lead to completely different conclusions about the population data. Thus, it is necessary to look for these too large or small numerical values and it has to be decided whether to omit or include these values in calculating the mean.

Weighted mean is calculated sometimes where certain weights are to be associated to all the observations.

Median

Median is the central value of an ordered set or the average of the middle two values if the data set size is even. Compared to the mean, median has certain advantages like it does not get influenced by too small or too big values. Median for the speed data, after arranging the data in order, is 59 kmph. Median for the compressive strength data, (after arranging the data in order and the average of the middle two values), is 32.665 MPa.

Table: 28 day Compressive strength of 108 cubes of M-25 grade concrete

37.33	32.22	32.89	36	37.11	33.33
37.56	32	32.89	36.22	37.33	33.56
37.56	32	32.89	36.44	37.33	33.56
34.89	31.33	26.22	34.44	34.89	32.22
34.89	31.11	26.22	34.67	35.11	32.44
35.11	30.89	26.44	34.89	35.11	32
31.78	32.44	28.89	32	32.33	35.11
31.78	32.44	28.67	31.76	32	35.33
32	32.22	29.11	31.76	32	32.33
34.89	33.56	31.78	34.22	34.67	34
35.11	33.56	32	34.67	35.11	33.78
35.11	33.56	32	34.67	34.67	33.78
32.22	35.56	31.56	32	37.33	24
32.44	35.56	31.78	32	37.56	24
32.44	35.11	31.78	32	37.56	24.44
30.89	35.56	26.67	29.33	32.22	41.78
31.11	35.11	26.22	29.56	32.44	42.22
31.33	35.11	26.22	29.78	32.44	42.22

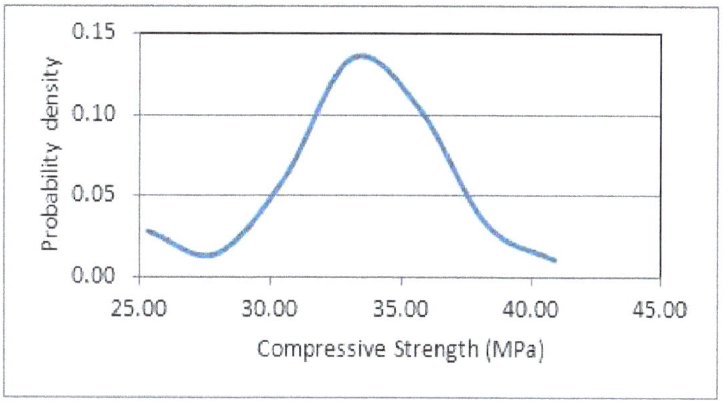

Probability density function of the compressive strength data

Mode

Mode is the most repeated value in the data set. For the speed data given in the table, the mode is 55km/hr. Similarly, the mode for the compressive strength data is 32 MPa. Compared to the mean and median, mode has several advantages. It has the highest probability of occurrence and is not influenced by small or big values.

Measures of Dispersion

Measures of dispersion are useful in studying the variability or degree of the scatter observed in the data. In the absence of too small or too big values *range* is one useful descriptor for the dispersion.

Standard Deviation

Standard deviation is the root mean square deviation about the mean. Standard deviation is the principal measure of the dispersion of the data. In case of the sample data it is denoted with s and in case of population data it is denoted with σ. $\hat{s}$ is the estimator for the population standard deviation in the absence of σ. Standard deviation of the sample is given by;

$s=\sqrt{\frac{1}{n-1}\sum_{i=1}^{n}\left(x_i-\bar{x}\right)^2}$, where, $\bar{x}$ is the sample mean and n-1 is the degrees of freedom. One way to explain the degrees of freedom is that, n-1 deviations are enough to compute the summation in the above equation. The other way to understand, it is the number of independent deviations, about the sample mean, available in estimating the sample standard deviation. Since we know that;

$$\bar{x}=\frac{1}{n}\sum_{i=1}^{n}x_i$$

only n-1 deviations about the sample mean are independent and based on the above equation the remaining deviation is known. Based on the above formula for the mean, the n th deviation about the mean is nothing but;

$$x_n-\bar{x}=-\sum_{i=1}^{n-1}\left(x_i-\bar{x}\right)$$

Sample standard deviation shown above is only an unbiased estimate of the population standard deviation.

Standard deviation of the population is computed as given below;

$\sigma=\sqrt{\frac{1}{N}\sum_{i=1}^{n}\left(x_i-\mu\right)^2}$, where μ is the population mean and N is the number of data points.

In the absence of the population standard deviation it is estimated using the sample data and is as given below. $\hat{s}$ is an unbiased estimator for the population standard deviation. If the sample size n is used in place of n-1 the estimate is going to be a biased estimate.

$$\hat{s}=\sqrt{\frac{1}{n-1}\sum_{i=1}^{n}\left(x_i-\bar{x}\right)^2}.$$

Coefficient of Variation

This is a dimensionless measure of dispersion and it is defined as the ratio of the standard deviation and mean. It is defined as follows;

$$\upsilon=\frac{s}{\bar{x}}$$

This measure is useful in comparing different data sets with respect to the central tendency and dispersion.

Variance

Variance is one of the important measures of dispersion and if its value is high the variability in the data is high and if it is less the variability in the data is less.

$$\sigma^2 = \frac{1}{n}\sum_{i=1}^{n}(x_i - \mu)^2$$

Similar to the population standard deviation, for population variance also an estimate can be obtained using the sample data.

Grouped Data

When dealing with a large set of data, it is necessary to group the data into different classes and the formulae used for computing the mean and the standard deviation are different.

If the data is grouped into c different classes and f_j is the frequency of the data in the jth class, and x_j is the midpoint of the class then the mean is computed as follows;

$$\overline{x} = \frac{\sum_{j=1}^{c} f_j x_j}{\sum_{j=1}^{c} f_j}$$

Variance is computed using the frequency data as follows;

$$s^2 = \frac{\sum_{j=1}^{c} f_j x_j^2 - \frac{\left\{\sum_{j=1}^{c} f_j x_j\right\}^2}{n}}{n-1}$$

Where c is number of classes and f_j is the frequency in the jth interval.

Median for the Grouped Data

Median for the grouped data can be obtained in the following way;

1. Get the cumulative frequencies and find out the class in which the median is falling. This is known as the median class. Absolute frequency corresponding to this class is termed as the frequency of median class, and it is denoted with f_m.

2. Cumulative frequency up to the class before the median class is denoted with the letter F.

3. The median can be computed as follows;

$$Median = L_m + \left(\frac{\frac{N}{2} - F}{f_m} \right) * c$$

Where, L_m is the lower limit of the median class, c is the class width, and N is the total frequency.

Mode for the Grouped Data

If the data are not grouped, as discussed earlier, the mode is the value which has the highest frequency in the data set. In case of grouped data the modal class is the class with the highest frequency. For grouped data the mode is computed as described below;

$$Mode = L_{\text{mode}} + \left(\frac{\Delta_1}{\Delta_1 + \Delta_2} \right) * c$$

Where, L_{mode} is the lower limit of the modal class, Δ_1 is the frequency difference between the modal class and the next class, Δ_2 is the frequency difference between the modal class and the class before the modal class.

Example

For the speed data given in the table, the mean and the standard deviation can be found by dividing the data into several classes of equal length. The calculations are shown in the table.

Mean of the frequency data is,

$$\begin{aligned} \overline{x} &= \frac{\sum_{j=1}^{c} f_j x_j}{\sum_{j=1}^{c} f_j} \\ &= \frac{10587}{169} \\ &= 62.64 \, or \, 63 \end{aligned}$$

Similarly, the standard deviation is,

$$\begin{aligned} s^2 &= \frac{\sum_{j=1}^{c} f_j x_j^2 - \frac{\left\{\sum_{j=1}^{c} f_j x_j\right\}^2}{n}}{n-1} \\ &= \frac{680751 - 663222}{168} \\ &= 104.34 \end{aligned}$$

Table: Calculations for computing the mean and standard deviation for the grouped data on speed

Class lower limit	Class upper limit	Class middle value (x_i)	Absolute frequency (f_i)	Cumulative frequency	$f_i^* x_i$	$f_i^* x_i^2$
36	40	38	2	2	76	2888
41	45	43	5	7	215	9245
46	50	48	14	21	672	32256
51	55	53	19	40	1007	53371
56	60	58	31	71	1798	104284
61	65	63	33	104	2079	130977
66	70	68	31	135	2108	143344
71	75	73	14	149	1022	74606
76	80	78	11	160	858	66924
81	85	83	8	168	664	55112
86	90	88	1	169	88	7744

Median

$$Median = L_m + \left(\frac{\frac{N}{2} - F}{f_m} \right) * c$$

From the table above the following information can be obtained;

Lower limit of the median class, L_m : 61

N = 169

Frequency up to the median class, F: 71

Frequency corresponding to the median class, f_m : 33

Class width, c: 4

$$Median = 61 + \left(\frac{\frac{169}{2} - 71}{33} \right) * 4$$

$$= 61 + 1.64$$

$$= 62.64 \text{ or } 63$$

Mode

$$Mode = L_{\text{mode}} + \left(\frac{\Delta_1}{\Delta_1 + \Delta_2} \right) * c$$

Lower limit of the mode class, L_{mode} : 61

Frequency difference between the modal class and the next class, Δ_1: 2

Frequency difference between the modal class and the previous class, Δ_2: 2

Class width, c: 4

$$Mode = 61 + \left(\frac{2}{4}\right) * 4$$
$$= 63$$

It can be seen that the all the three values (mean, mode, and median for the speed data) are equal to 63, hence this distribution can be called as the symmetric distribution.

For Compressive Strength Data

Table: Calculations for computing the mean, mode, and median for the grouped data on compressive strength of concrete

Lower Limit	Upper Limit	Middle Value (x_i)	Frequency (f_i)	Cumulative Frequency	$f_i^* x_i^2$
24.00	26.60	25.30	8	8	202.41
20.00	29.20	27.90	4	12	111.02
29.21	31.81	30.51	17	29	518.62
31.81	34.41	33.11	38	67	1258.18
34.41	37.01	35.71	29	96	1035.67
37.02	39.62	38.32	9	105	344.84
39.62	42.22	40.92	3	108	122.76

Mean

$$\bar{x} = \frac{\sum_{j=1}^{c} f_j x_j}{\sum_{j=1}^{c} f_j}$$
$$= \frac{3594.1}{108}$$
$$= 33.28\ km / Hr$$

Median

$$Median = L_m + \left(\frac{\frac{N}{2} - F}{f_m}\right) * c$$

From the table, following information can be obtained;

Lower limit of the median class, L_m : 31.81

N = 108

Frequency up to the median class, F: 29

Frequency corresponding to the median class, f_m : 38

Class width, c: 2.6

$$Median = 31.81 + \left(\frac{\frac{108}{2} - 29}{38} \right) * 2.6$$

$$= 31.81 + 1.71$$

$$33.52$$

Mode

$$Mode = L_{Mode} + \left(\frac{\Delta_1}{\Delta_1 + \Delta_2} \right) * c$$

Lower limit of the mode class, L_{mode} : 31.81

Frequency difference between the modal class and the next class, Δ_1: 9

Frequency difference between the modal class and the previous class, Δ_2: 21

Class width, c: 4

$$Mode = 31.81 + \left(\frac{9}{30} \right) * 2.6$$

$$= 31.81 + 0.78$$

$$= 32.59$$

Based on the values of mean, mode, median it can be said that the distribution of the compressive strength data are also approximately symmetrical.

References

- Griggs, Francis E Jr. "Amos Eaton was Right!". Journal of Professional Issues in Engineering Education and Practice, Vol. 123, No. 1, January 1997, pp. 30–34
- Oakes, William C.; Leone, Les L.; Gunn, Craig J. (2001). Engineering Your Future. Great Lakes Press. ISBN 1-881018-57-1

- Nancy R. Tague (2004). "Seven Basic Quality Tools". The Quality Toolbox. Milwaukee, Wisconsin: American Society Quality. p. 15. Retrieved 2010-02-05
- Sturges, H. A. (1926). "The choice of a class interval". Journal of the American Statistical Association: 65–66. doi:10.1080/01621459.1926.10502161. JSTOR 2965501
- Charles Stangor (2011) "Research Methods For The Behavioral Sciences". Wadsworth, Cengage Learning. ISBN 9780840031976

2

Probability Concepts and Random Events

Probability is the study of chance and events that may take place or not. It calculates the possibility of an event to happen, with the results falling between zero and one. The closer the evaluation of an event is to one, the higher is its chances to occur. This chapter elucidates the crucial theories and principles of probability.

Probability

Probability is the measure of the likelihood that an event will occur. Probability is quantified as a number between 0 and 1 (where 0 indicates impossibility and 1 indicates certainty). The higher the probability of an event, the more certain that the event will occur. A simple example is the tossing of a fair (unbiased) coin. Since the coin is unbiased, the two outcomes ("head" and "tail") are both equally probable; the probability of "head" equals the probability of "tail". Since no other outcomes are possible, the probability is 1/2 (or 50%), of either "head" or "tail". In other words, the probability of "head" is 1 out of 2 outcomes and the probability of "tail" is also 1 out of 2 outcomes, expressed as 0.5 when converted to decimal, with the above-mentioned quantification system. This type of probability is also called a priori probability.

These concepts have been given an axiomatic mathematical formalization in probability theory, which is used widely in such areas of study as mathematics, statistics, finance, gambling, science (in particular physics), artificial intelligence/machine learning, computer science, game theory, and philosophy to, for example, draw inferences about the expected frequency of events. Probability theory is also used to describe the underlying mechanics and regularities of complex systems.

Interpretations

When dealing with experiments that are random and well-defined in a purely theoretical setting (like tossing a fair coin), probabilities can be numerically described by the number of desired outcomes divided by the total number of all outcomes. For example, tossing a fair coin twice will yield "head-head", "head-tail", "tail-head", and "tail-tail" outcomes. The probability of getting an outcome of "head-head" is 1 out of 4 outcomes or 1/4 or 0.25 (or 25%). When it comes to practical application however, there are two major competing categories of probability interpretations, whose adherents possess different views about the fundamental nature of probability:

1. Objectivists assign numbers to describe some objective or physical state of affairs. The most popular version of objective probability is frequentist probability, which claims that the probability of a random event denotes the *relative frequency of occurrence* of an experiment's outcome, when repeating the experiment. This interpretation considers probability to be the relative frequency "in the long run" of outcomes. A modification of this is

propensity probability, which interprets probability as the tendency of some experiment to yield a certain outcome, even if it is performed only once.

2. Subjectivists assign numbers per subjective probability, i.e., as a degree of belief. The degree of belief has been interpreted as, "the price at which you would buy or sell a bet that pays 1 unit of utility if E, 0 if not E." The most popular version of subjective probability is Bayesian probability, which includes expert knowledge as well as experimental data to produce probabilities. The expert knowledge is represented by some (subjective) prior probability distribution. These data are incorporated in a likelihood function. The product of the prior and the likelihood, normalized, results in a posterior probability distribution that incorporates all the information known to date. Starting from arbitrary, subjective probabilities for a group of agents, some Bayesians claim that all agents will eventually have sufficiently similar assessments of probabilities, given enough evidence.

Etymology

The word *probability* derives from the Latin *probabilitas*, which can also mean "probity", a measure of the authority of a witness in a legal case in Europe, and often correlated with the witness's nobility. In a sense, this differs much from the modern meaning of *probability*, which, in contrast, is a measure of the weight of empirical evidence, and is arrived at from inductive reasoning and statistical inference.

History

The scientific study of probability is a modern development of mathematics. Gambling shows that there has been an interest in quantifying the ideas of probability for millennia, but exact mathematical descriptions arose much later. There are reasons of course, for the slow development of the mathematics of probability. Whereas games of chance provided the impetus for the mathematical study of probability, fundamental issues are still obscured by the superstitions of gamblers.

Christiaan Huygens probably published the first book on probability

According to Richard Jeffrey, "Before the middle of the seventeenth century, the term 'probable'

(Latin *probabilis*) meant *approvable*, and was applied in that sense, univocally, to opinion and to action. A probable action or opinion was one such as sensible people would undertake or hold, in the circumstances." However, in legal contexts especially, 'probable' could also apply to propositions for which there was good evidence.

Gerolamo Cardano

The sixteenth century Italian polymath Gerolamo Cardano demonstrated the efficacy of defining odds as the ratio of favourable to unfavourable outcomes (which implies that the probability of an event is given by the ratio of favourable outcomes to the total number of possible outcomes). Aside from the elementary work by Cardano, the doctrine of probabilities dates to the correspondence of Pierre de Fermat and Blaise Pascal (1654). Christiaan Huygens (1657) gave the earliest known scientific treatment of the subject. Jakob Bernoulli's *Ars Conjectandi* (posthumous, 1713) and Abraham de Moivre's *Doctrine of Chances* (1718) treated the subject as a branch of mathematics.

The theory of errors may be traced back to Roger Cotes's *Opera Miscellanea* (posthumous, 1722), but a memoir prepared by Thomas Simpson in 1755 (printed 1756) first applied the theory to the discussion of errors of observation. The reprint (1757) of this memoir lays down the axioms that positive and negative errors are equally probable, and that certain assignable limits define the range of all errors. Simpson also discusses continuous errors and describes a probability curve.

The first two laws of error that were proposed both originated with Pierre-Simon Laplace. The first law was published in 1774 and stated that the frequency of an error could be expressed as an exponential function of the numerical magnitude of the error, disregarding sign. The second law of error was proposed in 1778 by Laplace and stated that the frequency of the error is an exponential function of the square of the error. The second law of error is called the normal distribution or the Gauss law. "It is difficult historically to attribute that law to Gauss, who in spite of his well-known precocity had probably not made this discovery before he was two years old."

Daniel Bernoulli (1778) introduced the principle of the maximum product of the probabilities of a system of concurrent errors.

Carl Friedrich Gauss

Adrien-Marie Legendre (1805) developed the method of least squares, and introduced it in his *Nouvelles méthodes pour la détermination des orbites des comètes* (*New Methods for Determining the Orbits of Comets*). In ignorance of Legendre's contribution, an Irish-American writer, Robert Adrain, editor of "The Analyst" (1808), first deduced the law of facility of error,

$$\phi(x) = ce^{-h^2x^2},$$

where h is a constant depending on precision of observation, and c is a scale factor ensuring that the area under the curve equals 1. He gave two proofs, the second being essentially the same as John Herschel's (1850). Gauss gave the first proof that seems to have been known in Europe (the third after Adrain's) in 1809. Further proofs were given by Laplace (1810, 1812), Gauss (1823), James Ivory (1825, 1826), Hagen (1837), Friedrich Bessel (1838), W. F. Donkin (1844, 1856), and Morgan Crofton (1870). Other contributors were Ellis (1844), De Morgan (1864), Glaisher (1872), and Giovanni Schiaparelli (1875). Peters's (1856) formula for r, the probable error of a single observation, is well known.

In the nineteenth century authors on the general theory included Laplace, Sylvestre Lacroix (1816), Littrow (1833), Adolphe Quetelet (1853), Richard Dedekind (1860), Helmert (1872), Hermann Laurent (1873), Liagre, Didion, and Karl Pearson. Augustus De Morgan and George Boole improved the exposition of the theory.

Andrey Markov introduced the notion of Markov chains (1906), which played an important role in stochastic processes theory and its applications. The modern theory of probability based on the measure theory was developed by Andrey Kolmogorov (1931).

On the geometric side contributors to *The Educational Times* were influential (Miller, Crofton, McColl, Wolstenholme, Watson, and Artemas Martin).

Theory

Like other theories, the theory of probability is a representation of probabilistic concepts in formal terms—that is, in terms that can be considered separately from their meaning. These formal terms are manipulated by the rules of mathematics and logic, and any results are interpreted or translated back into the problem domain.

There have been at least two successful attempts to formalize probability, namely the Kolmogorov formulation and the Cox formulation. In Kolmogorov's formulation, sets are interpreted as events and probability itself as a measure on a class of sets. In Cox's theorem, probability is taken as a primitive (that is, not further analyzed) and the emphasis is on constructing a consistent assignment of probability values to propositions. In both cases, the laws of probability are the same, except for technical details.

There are other methods for quantifying uncertainty, such as the Dempster–Shafer theory or possibility theory, but those are essentially different and not compatible with the laws of probability as usually understood.

Applications

Probability theory is applied in everyday life in risk assessment and modeling. The insurance industry and markets use actuarial science to determine pricing and make trading decisions. Governments apply probabilistic methods in environmental regulation, entitlement analysis (Reliability theory of aging and longevity), and financial regulation.

A good example of the use of probability theory in equity trading is the effect of the perceived probability of any widespread Middle East conflict on oil prices, which have ripple effects in the economy as a whole. An assessment by a commodity trader that a war is more likely can send that commodity's prices up or down, and signals other traders of that opinion. Accordingly, the probabilities are neither assessed independently nor necessarily very rationally. The theory of behavioral finance emerged to describe the effect of such groupthink on pricing, on policy, and on peace and conflict.

In addition to financial assessment, probability can be used to analyze trends in biology (e.g. disease spread) as well as ecology (e.g. biological Punnett squares). As with finance, risk assessment can be used as a statistical tool to calculate the likelihood of undesirable events occurring and can assist with implementing protocols to avoid encountering such circumstances. Probability is used to design games of chance so that casinos can make a guaranteed profit, yet provide payouts to players that are frequent enough to encourage continued play.

The discovery of rigorous methods to assess and combine probability assessments has changed society. It is important for most citizens to understand how probability assessments are made, and how they contribute to decisions.

Another significant application of probability theory in everyday life is reliability. Many consumer products, such as automobiles and consumer electronics, use reliability theory in product design to reduce the probability of failure. Failure probability may influence a manufacturer's decisions on a product's warranty.

The cache language model and other statistical language models that are used in natural language processing are also examples of applications of probability theory.

Mathematical Treatment

Consider an experiment that can produce a number of results. The collection of all possible results is called the sample space of the experiment. The power set of the sample space is formed by considering all different collections of possible results. For example, rolling a dice can produce six possible results. One collection of possible results gives an odd number on the dice. Thus, the subset {1,3,5} is an element of the power set of the sample space of dice rolls. These collections are called "events". In this case, {1,3,5} is the event that the dice falls on some odd number. If the results that actually occur fall in a given event, the event is said to have occurred.

A probability is a way of assigning every event a value between zero and one, with the requirement that the event made up of all possible results (in our example, the event {1,2,3,4,5,6}) is assigned a value of one. To qualify as a probability, the assignment of values must satisfy the requirement that if you look at a collection of mutually exclusive events (events with no common results, e.g., the events {1,6}, {3}, and {2,4} are all mutually exclusive), the probability that at least one of the events will occur is given by the sum of the probabilities of all the individual events.

The probability of an event A is written as $P(A)$, $p(A)$, or $\Pr(A)$. This mathematical definition of probability can extend to infinite sample spaces, and even uncountable sample spaces, using the concept of a measure.

The *opposite* or *complement* of an event A is the event [not A] (that is, the event of A not occurring), often denoted as $\bar{A}, A^{C}, \neg A$, or $\sim A$; its probability is given by $P(\text{not } A) = 1 - P(A)$. As an example, the chance of not rolling a six on a six-sided die is 1 – (chance of rolling a six) $= 1 - \frac{1}{6} = \frac{5}{6}$.

If two events A and B occur on a single performance of an experiment, this is called the intersection or joint probability of A and B, denoted as $P(A \cap B)$.

Independent Events

If two events, A and B are independent then the joint probability is

$$P(A \text{ and } B) = P(A \cap B) = P(A)P(B),$$

for example, if two coins are flipped the chance of both being heads is $\frac{1}{2} \times \frac{1}{2} = \frac{1}{4}$.

Mutually Exclusive Events

If either event A or event B occurs on a single performance of an experiment this is called the union of the events A and B denoted as $P(A \cup B)$. If two events are mutually exclusive then the probability of either occurring is

$$P(A \text{ or } B) = P(A \cup B) = P(A) + P(B).$$

For example, the chance of rolling a 1 or 2 on a six-sided die is $P(1 \text{ or } 2) = P(1) + P(2) = \frac{1}{6} + \frac{1}{6} = \frac{1}{3}$.

Not mutually Exclusive Events

If the events are not mutually exclusive then

$$P(A \text{ or } B) = P(A) + P(B) - P(A \text{ and } B).$$

For example, when drawing a single card at random from a regular deck of cards, the chance of getting a heart or a face card (J,Q,K) (or one that is both) is $\frac{13}{52} + \frac{12}{52} - \frac{3}{52} = \frac{11}{26}$, because of the 52 cards of a deck 13 are hearts, 12 are face cards, and 3 are both: here the possibilities included in the "3 that are both" are included in each of the "13 hearts" and the "12 face cards" but should only be counted once.

Conditional Probability

Conditional probability is the probability of some event *A*, given the occurrence of some other event *B*. Conditional probability is written $P(A|B)$ and is read "the probability of *A*, given *B*". It is defined by

$$P(A|B) = \frac{P(A \cap B)}{P(B)}.$$

If $P(B) = 0$ then $P(A|B)$ is formally undefined by this expression. However, it is possible to define a conditional probability for some zero-probability events using a σ-algebra of such events (such as those arising from a continuous random variable).

For example, in a bag of 2 red balls and 2 blue balls (4 balls in total), the probability of taking a red ball is $1/2$; however, when taking a second ball, the probability of it being either a red ball or a blue ball depends on the ball previously taken, such as, if a red ball was taken, the probability of picking a red ball again would be $1/3$ since only 1 red and 2 blue balls would have been remaining.

Inverse Probability

In probability theory and applications, Bayes' rule relates the odds of event A_1 to event A_2, before (prior to) and after (posterior to) conditioning on another event B. The odds on A_1 to event A_2 is simply the ratio of the probabilities of the two events. When arbitrarily many events A are of interest, not just two, the rule can be rephrased as posterior is proportional to prior times likelihood, $P(A \mid B) \propto P(A)P(B \mid A)$ where the proportionality symbol means that the left hand side is proportional to (i.e., equals a constant times) the right hand side as A varies, for fixed or given B. (Lee, 2012; Bertsch McGrayne, 2012). In this form it goes back to Laplace (1774) and to Cournot (1843).

Summary of probabilities	
Event	**Probability**
A	$P(A) \in [0,1]$
not A	$P(A^C) = 1 - P(A)$
A or B	$P(A \cup B) = P(A) + P(B) - P(A \cap B)$ $P(A \cup B) = P(A) + P(B)$ if A and B are mutually exclusive
A and B	$P(A \cap B) = P(A \mid B)P(B) = P(B \mid A)P(A)$ $P(A \cap B) = P(A)P(B)$ if A and B are independent
A given B	$P(A \mid B) = \frac{P(A \cap B)}{P(B)} = \frac{P(B \mid A)P(A)}{P(B)}$

Relation to Randomness and Probability in Quantum Mechanics

In a deterministic universe, based on Newtonian concepts, there would be no probability if all conditions were known (Laplace's demon), (but there are situations in which sensitivity to initial conditions exceeds our ability to measure them, i.e. know them). In the case of a roulette wheel, if the force of the hand and the period of that force are known, the number on which the ball will stop would be a certainty (though as a practical matter, this would likely be true only of a roulette wheel that had not been exactly levelled – as Thomas A. Bass' Newtonian Casino revealed). Of course, this also assumes knowledge of inertia and friction of the wheel, weight, smoothness and roundness of the ball, variations in hand speed during the turning and so forth. A probabilistic description can thus be more useful than Newtonian mechanics for analyzing the pattern of outcomes of repeated rolls of a roulette wheel. Physicists face the same situation in kinetic theory of gases, where the system, while deterministic *in principle*, is so complex (with the number of molecules typically the order of magnitude of Avogadro constant 6.02×10^{23}) that only a statistical description of its properties is feasible.

Probability theory is required to describe quantum phenomena. A revolutionary discovery of early 20th century physics was the random character of all physical processes that occur at sub-atomic scales and are governed by the laws of quantum mechanics. The objective wave function evolves deterministically but, according to the Copenhagen interpretation, it deals with probabilities of observing, the outcome being explained by a wave function collapse when an observation is made. However, the loss of determinism for the sake of instrumentalism did not meet with universal approval. Albert Einstein famously remarked in a letter to Max Born: "I am convinced that God does not play dice". Like Einstein, Erwin Schrödinger, who discovered the wave function, believed quantum mechanics is a statistical approximation of an underlying deterministic reality. In some modern interpretations of the statistical mechanics of measurement, quantum decoherence is invoked to account for the appearance of subjectively probabilistic experimental outcomes.

The term probability is related to the degree of certainty about a particular event of interest. For example, in designing the cross drainage structures the most probabilistic maximum water level of the water body is a very important parameter. Knowledge of probability is essential in knowing this most probabilistic water level that may occur in that water body. Before going into the theory of probability it is important to have some understanding of the theory of sets.

Sample Space and Event

Sample space , Ω, is the collection of the all possible outcomes related to the set of events associated to any random experiment. The term sample is attached to the space because of the uncertainty associated to the experiment. If a coin is tossed once the resulting sample space is, Ω = {H, T}, where H corresponds to the *heads* and T to the *tails* . If a coin is tossed twice, the experiment is tossing the coin twice and the resulting sample space is, Ω = {HH, HT, TH, TT}. Cardinality of the sample space is nothing but the number of elements in the sample space. In the example 'tossing the coin once' the cardinality of the sample space, $|\Omega|$, is 2. Sample space can be either discrete or continuous depending on the experiment being dealt with. For example the sample space is continuous when the experiment under consideration is the speed measurements at a road section or the water level measurements in a reservoir. Similarly the sample space is discrete if the experiment under consideration is the occurrence of number of accidents over a unit road length or the number of days in a month with rainfall greater than certain value.

The outcomes associated to an *event* are a subset of the sample space Ω . An *event* associated with a random experiment corresponds to a question about the experiment that has yes or no answer, and this in turn is associated with a subset of the sample space. For example, in the experiment 'tossing the coin twice', question saying 'is it heads in the first toss?' is an event and the sample space associated to the answer 'yes' is A= {HH, HT}. Event space is denoted with A. The event space A is a special set that contains the sample space. The event space is characterized by the following properties:

Event space is the collection of all possible events associated to a given experiment. The event space contains all the elements of the sample space as well as the elements corresponding to the events defined as the combinations of the events related to that experiment. Cardinality of the event space is 2n, where, n is the number of elements in the corresponding sample space. Event space is denoted with A. The event space A is a special set that contains the sample space. The event space is characterized by the following properties:

1. $\Omega \in A$
2. If $A \in A$ then $A^c \in A$
3. If $A_1 \in A$ and $A_2 \in A$, then $A_1 + A_2 \in A$

There are several examples from the Civil Engineering perspective. In case of a reservoir, water levels vary with respect to the seasons. During some period of the year water level may be just the dead storage level and during another period it may be above the supply storage levels and these fluctuations are random and the corresponding random variable is continuous in nature. The sample space corresponding to this random variable, in discrete form, can be Ω = {S: 0≤S<h}, where the 0 is referring to the dead storage level and the h is referring to the maximum flood control storage level. In the sample space, as shown in the following figure, A can be an event representing the flow levels in supply storage levels.

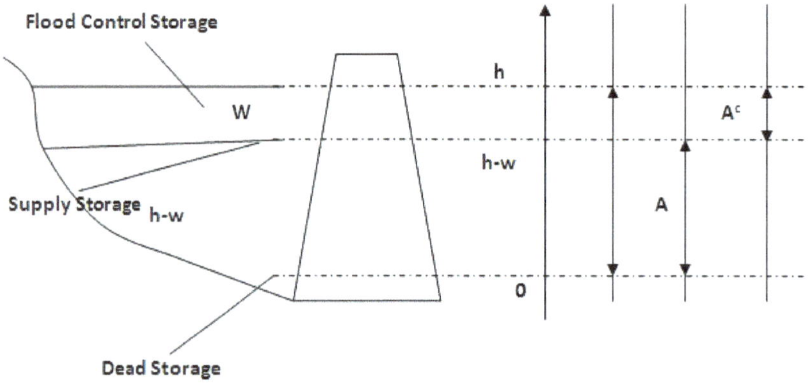

Schematic representation of the water storage levels in a dam

Countable Sample Space

If the sample space is discrete then it is also called as the countable sample space because the sample space is countably finite or infinite. If the sample space is countable then the event space is also countable. The annual number of accidents occurring on a road stretch of unit length is countably finite. Similarly the number of vehicles crossing a section of the road in a unit time period is countably finite. The time gap between earthquakes of certain magnitude expressed in number of days is countably infinite.

Uncountable Sample Space

If the sample space is continuous then it is also called as the uncountable sample space. In this case the event space is also uncountable. For the speed example, the event space for the speed of the n^{th}vehicle crossing the road section is uncountable. The compressive strength of concrete, the CBR value of the sub-grade, or the BOD of a particular water sample are some other examples for the uncountable sample space.

Different Approaches to Probability

Various approaches have been used to define the probability and few approaches among them have certain drawbacks when using them in real life applications. Broadly they can be divided into two categories namely objective, and subjective. The objective approaches rely on the priori knowledge of the experiment. The subjective approaches also rely on the priori knowledge of the experiment but it is subjective to the individual dealing with the experiment. Some individuals may have full knowledge of the experiment and some may have partial knowledge. The following sections elaborate three approaches to probability namely classical, relative frequency, and axiomatic.

Classical Approach

If an event A occurs in m times out of a total of n exhaustive, mutually exclusive and equally likely experiments, then the probability of the occurrence of event A is m/n .

$$p(A) = m/n$$

The sample space corresponding to the experiment 'tossing a fair coin' is shown in the figure below.

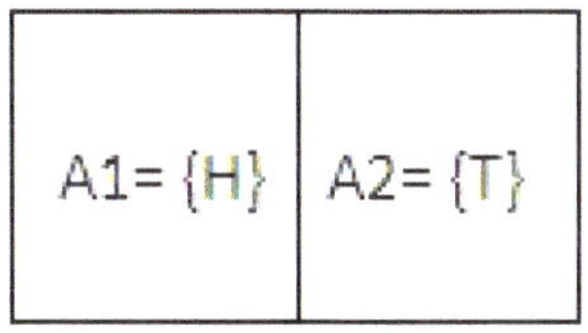

Sample space, **Ω,** for the experiment 'tossing a fair coin'

In this case, the two possible events are getting heads and getting tails. These two events are collectively exhaustive because all the elements in the sample space, shown in the above Figure, are falling in either of the two possible events. They are mutually exclusive because there is no overlap in the sample space corresponding two the two possible events.

To apply this definition one must have prior knowledge on the sample space associated to the experiment and the sample space associated to the event. For example, what is the probability of getting a card of a particular color when one card is randomly picked from a pack of cards? Since the information on the color composition of cards is known (i.e. m and n are known to be 26 and 52 respectively), answer to this question is 0.5.

Relative Frequency Approach

Similarly, this definition can also be used in situations where the experiments can be conducted in controlled environments and the ratio of favorable outcomes to an event and the total repetitions of the experiment, gives the probability. This is known as the relative frequency and n must be large to get the realistic probability value associated to this event.

$$p(A) = \lim_{n\to\infty} \frac{m}{n}$$

This approach has certain disadvantages such as the probability is never irrational i.e. always it takes rational real numbers. This definition cannot be used in situation where the exhaustive number of experiments cannot be conducted and nothing can be said about whether they are equally likely even if it is possible to conduct the experiments. Examples are what is the probability of having certain amount of rainfall in a given month, what is the most severe wind load that is going to act on a tall structure.

Axiomatic Approach

An axiomatic approach is taken to define probability as a set function where the elements of the domain are the sets and the elements of range are real numbers. If event A is an element in the domain of this function, P(A) is the customary notation used to designate the corresponding element in the range.

Probability Function

A probability function p(A) is a function mapping the event space A of a random experiment into the interval [0,1] y of A given B is defined as the quotient of the probability of the joint of events A and B, and the probability of B:according to the following axioms;

- Axiom 1. For any event A, $0 \leq P(A) \leq 1$
- Axiom 2. $P(\Omega) = 1$
- Axiom 3. If A and B are any two mutually exclusive events then,

$P(A \cup B)) = P(A) + P(B)$

As given in the third axiom the addition property of the probability can be extended to any number of events as long as the events are mutually exclusive. If the events are not mutually exclusive then;

$P(A \cup B) = P(A) + P(B) - P(A \cap B)$

$P(A \cap B)$ is Φ if both the events are mutually exclusive.

Problem

In designing the right turning lanes at intersections it is necessary to know the probable number of cars that are taking right turn. A simple T-junction with a separate right turning lane is shown in the following figure.

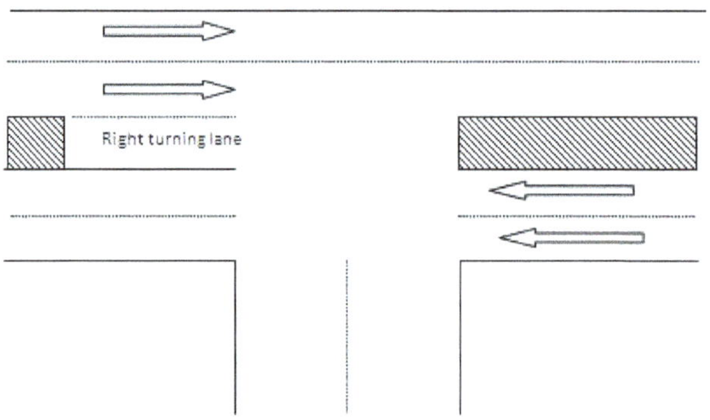

Schematic diagram of the three legged intersection

From the observations made on the right turning vehicles the following data has been obtained.

Table: Frequency calculations of the right turning vehicles near a junction.

No. of vehicles	Frequency	Relative frequency
0	4	4/60
1	13	13/60
2	21	21/60
3	14	14/60
4	4	4/60
5	3	3/60
6	1	1/60
7	0	0
8	0	0

If the events are defined as follows;

E_1 = More than 3 vehicles waiting for the right turn

E_2 = at most 3 vehicles waiting for the right turn

The arrival of right turning vehicles are certainly independent events hence it is assumed that the number of vehicles waiting for right turning are mutually exclusive.

Probability associated to the event E_1 is

$p(E_1)$ = probability associated to the event "more than 3 vehicles waiting for right turn".

Relative frequencies may be used as an approximation for the associated probabilities.

$P(E_1) = 4/60 + 3/60 + 1/60 = 8/60$

Probability associated to the event E_2 is

$P(E_2)$ = probability associated to at most 3 vehicles waiting for the right turn

$= 4/60 + 13/60 + 21/60 + 14/60$

$= 52/60$

The events E_1 and E_2 are mutually exclusive and the corresponding probability of joint occurrence is zero. Union of these two events is equals to one.

$P(E_1 \cup E_2) = P(E_1) + p(E_2)$

$= 8/60 + 52/60$

$= 1.$

Conditional Probability

In probability theory, conditional probability is a measure of the probability of an event given that (by assumption, presumption, assertion or evidence) another event has occurred. If the event of interest is *A* and the event *B* is known or assumed to have occurred, "the conditional probability of *A* given *B*", or "the probability of *A* under the condition *B*", is usually written as $P(A|B)$, or sometimes $P_B(A)$. For example, the probability that any given person has a cough on any given day may be only 5%. But if we know or assume that the person has a cold, then they are much more likely to be coughing. The conditional probability of coughing given that you have a cold might be a much higher 75%.

The concept of conditional probability is one of the most fundamental and one of the most important concepts in probability theory. But conditional probabilities can be quite slippery and require careful interpretation. For example, there need not be a causal or temporal relationship between *A* and *B*.

$P(A|B)$ may or may not be equal to $P(A)$ (the unconditional probability of A). If $P(A|B) = P(A)$ (or its equivalent $P(B|A) = P(B)$), then events A and B are said to be independent: in such a case, having knowledge about either event does not change our knowledge about the other event. Also, in general, $P(A|B)$ (the conditional probability of A given B) is not equal to $P(B|A)$. For example, if you have cancer you might have a 90% chance of testing positive for cancer. In this case what is being measured is that the if event B "having cancer" has occurred, the probability A - *test is positive* given that B *having cancer* occurred is 90%, $P(A|B) = 90\%$. Alternatively, you can test positive for cancer but you may have only a 10% chance of actually having cancer because cancer is very rare. In this case what is being measured is the probability of the event B - *having cancer* given that the event A - *test is positive* has occurred, $P(B|A) = 10\%$. Falsely equating the two probabilities causes various errors of reasoning such as the base rate fallacy. Conditional probabilities can be correctly reversed using Bayes' theorem.

Definition

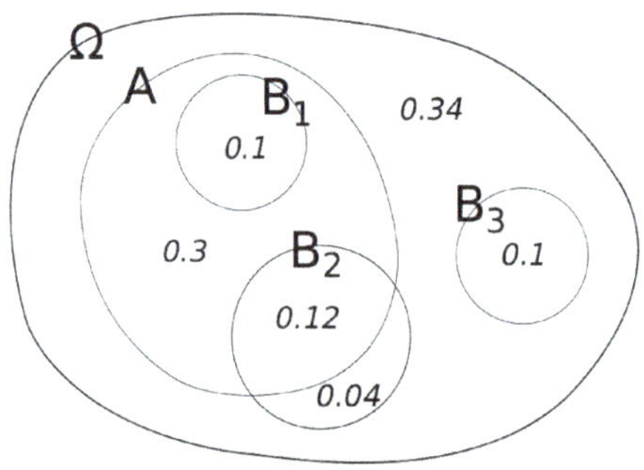

Illustration of conditional probabilities with an Euler diagram. The unconditional probability P(A) = 0.52. However, the conditional probability P(A|B1) = 1, P(A|B2) = 0.75, and P(A|B3) = 0.

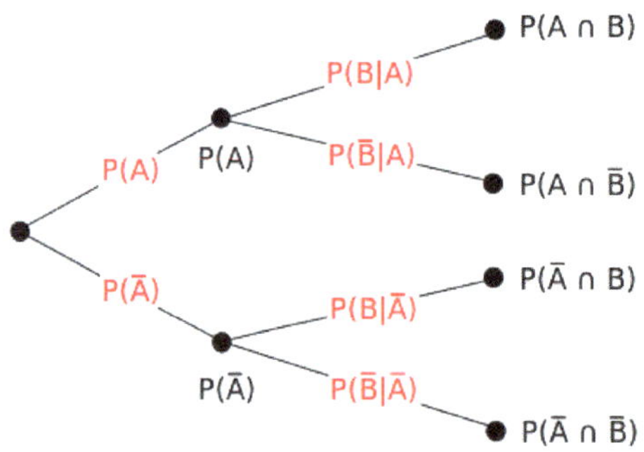

On a tree diagram, branch probabilities are conditional on the event associated with the parent node.

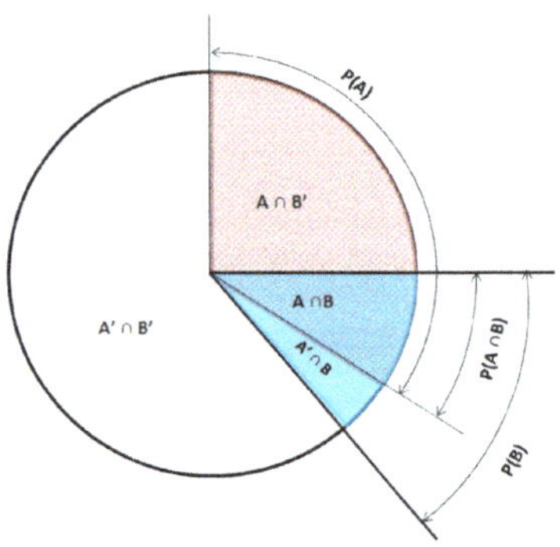

Venn Pie Chart describing conditional probabilities

Conditioning on an Event

Kolmogorov Definition

Given two events A and B from the sigma-field of a probability space with $P(B) > 0$, the conditional probability of A given B is defined as the quotient of the probability of the joint of events A and B, and the probability of B:

$$P(A \mid B) = \frac{P(A \cap B)}{P(B)}$$

This may be visualized as restricting the sample space to B. The logic behind this equation is that if the outcomes are restricted to B, this set serves as the new sample space.

Note that this is a definition but not a theoretical result. We just denote the quantity $P(A \cap B) / P(B)$ as $P(A \mid B)$ and call it the conditional probability of A given B.

As an Axiom of Probability

Some authors, such as De Finetti, prefer to introduce conditional probability as an axiom of probability:

$$P(A \cap B) = P(A \mid B)P(B)$$

Although mathematically equivalent, this may be preferred philosophically; under major probability interpretations such as the subjective theory, conditional probability is considered a primitive entity. Further, this "multiplication axiom" introduces a symmetry with the summation axiom for mutually exclusive events:

$$P(A \cup B) = P(A) + P(B) - \cancelto{0}{P(A \cap B)}$$

Measure-theoretic Definition

If $P(B) = 0$, then the simple definition of $P(A|B)$ is undefined. However, it is possible to define a conditional probability with respect to a σ-algebra of such events (such as those arising from a continuous random variable).

For example, if X and Y are non-degenerate and jointly continuous random variables with density $f_{X,Y}(x, y)$ then, if B has positive measure,

$$P(X \in A \mid Y \in B) = \frac{\int_{y \in B} \int_{x \in A} f_{X,Y}(x,y)\,dx\,dy}{\int_{y \in B} \int_{x \in \mathbb{R}} f_{X,Y}(x,y)\,dx\,dy}.$$

The case where B has zero measure is problematic. For the case that $B = \{y_0\}$, representing a single point, the conditional probability could be defined as

$$P(X \in \mathrm{A} | Y = y_0) = \frac{\int_{x \in A} f_{X,Y}(x, y_0) dx}{\int_{x \in \mathbb{R}} f_{X,Y}(x, y_0) dx},$$

however this approach leads to the Borel–Kolmogorov paradox. The more general case of zero measure is even more problematic, as can be seen by noting that the limit, as all δy_i approach zero, of

$$P(X \in A | Y \in \cup_i [y_i, y_i + \delta y_i]) \approxeq \frac{\sum_i \int_{x \in A} f_{X,Y}(x, y_i) dx \delta y_i}{\sum_i \int_{x \in \mathbb{R}} f_{X,Y}(x, y_i) dx \delta y_i},$$

Conditioning on a Random Variable

Conditioning on an event may be generalized to conditioning on a random variable. Let X be a random variable; we assume for the sake of presentation that X is discrete, that is, X takes on only finitely many values x. Let A be an event. The conditional probability of A given X is defined as the random variable, written $P(A|X)$, *that takes on the value*

$$P(A | X = x)$$

whenever

$$X = x.$$

More formally:

$$P(A \mid X)(\omega) = P(A | X = X(\omega)).$$

The conditional probability $P(A|X)$ is a function of X, e.g., if the function g is defined as

$$g(x) = P(A | X = x),$$

then

$$P(A \mid X) = g \circ X$$

Note that $P(A|X)$ and X are now both random variables. From the law of total probability, the expected value of $P(A|X)$ is equal to the unconditional probability of A.

Example

Suppose that somebody secretly rolls two fair six-sided dice, and we must predict the outcome (the sum of the two upward faces).

- Let A be the value rolled on die 1.

- Let B be the value rolled on die 2.

What is the probability that $A = 2$?

Table: shows the sample space of 36 outcomes

Clearly, $A = 2$ in exactly 6 of the 36 outcomes, thus $P(A=2) = 6/36 = 1/6$.

Table							
+		B					
		1	2	3	4	5	6
A	1	2	3	4	5	6	7
	2	3	4	5	6	7	8
	3	4	5	6	7	8	9
	4	5	6	7	8	9	10
	5	6	7	8	9	10	11
	6	7	8	9	10	11	12

Suppose it is revealed that $A+B \leq 5$

What is the probability $A+B \leq 5$?

Table: shows that $A+B \leq 5$ for exactly 10 of the same 36 outcomes, thus $P(A+B \leq 5) = 10/36$

+		B					
		1	2	3	4	5	6
A	1	2	3	4	5	6	7
	2	3	4	5	6	7	8
	3	4	5	6	7	8	9
	4	5	6	7	8	9	10
	5	6	7	8	9	10	11
	6	7	8	9	10	11	12

What is the probability that A = 2 given that A+B ≤ 5 ?

Table: shows that for 3 of these 10 outcomes, $A = 2$

Thus, the conditional probability $P(A=2 \mid A+B \leq 5) = 3/10 = 0.3$.

Table							
+		B					
		1	2	3	4	5	6
	1	2	3	4	5	6	7
	2	3	4	5	6	7	8

A	3	4	5	6	7	8	9
	4	5	6	7	8	9	10
	5	6	7	8	9	10	11
	6	7	8	9	10	11	12

Here in the earlier notation for the definition of conditional probability, the conditioning event B is that D1+D2 ≤ 5, and the event A is D1 = 2. We have $P(A \mid B) = \frac{P(A \cap B)}{P(B)} = \frac{3/36}{10/36} = \frac{3}{10}$, as seen in the table.

Use in Inference

In statistical inference, the conditional probability is an update of the probability of an event based on new information. Incorporating the new information can be done as follows

- Let *A* the event of interest be in the sample space, say (*X*,*P*).
- The occurrence of the event *A* knowing that event *B* has or will have occurred, means the occurrence of *A* as it is restricted to *B*, i.e. $A \cap B$.
- Without the knowledge of the occurrence of *B*, the information about the occurrence of *A* would simply be *P*(*A*)
- The probability of *A* knowing that event *B* has or will have occurred, will be the probability of $A \cap B$ compared with *P*(*B*), the probability *B* has occurred.
- This results in $P(A|B) = P(A \cap B)/P(B)$ whenever *P*(*B*)>0 and 0 otherwise.

Note: This approach results in a probability measure that is consistent with the original probability measure and satisfies all the Kolmogorov Axioms. This conditional probability measure also could have resulted by assuming that the relative magnitude of the probability of A with respect to X will be preserved with respect to B (cf. a Formal Derivation below).

The wording "evidence" or "information" is generally used in the Bayesian interpretation of probability. The conditioning event is interpreted as evidence for the conditioned event. That is, *P*(*A*) is the probability of *A* before accounting for evidence *E*, and *P*(*A*|*E*) is the probability of *A* after having accounted for evidence *E* or after having updated *P*(*A*). This is consistent with the frequentist interpretation, which is the first definition given above.

Statistical Independence

Events *A* and *B* are defined to be statistically independent if

$$P(A \cap B) = P(A)P(B).$$

If *P*(*B*) is not zero, then this is equivalent to the statement that

$$P(A \mid B) = P(A).$$

Similarly, if $P(A)$ is not zero, then

$$P(B \mid A) = P(B).$$

is also equivalent. Although the derived forms may seem more intuitive, they are not the preferred definition as the conditional probabilities may be undefined, and the preferred definition is symmetrical in A and B.

Important Distinctions: Independent Events vs Mutually Exclusive Events

The concepts of mutually independent events and mutually exclusive events are often confused.

As noted, statistical independence means:

$P(A|B) = P(A)$ and $P(B|A) = P(B)$, provided that the probability for the conditioning event is not zero.

However, mutually exclusive events implies that:

$P(A|B) = 0$, $P(B|A)=0$, and $P(A \cap B) = 0$.

Assuming Conditional Probability is of Similar Size to its Inverse

Relative size	Case B	Case $\bar{B}$	Total
Condition A	w	x	$w+x$
Condition $\bar{A}$	y	z	$y+z$
Total	$w+y$	$x+z$	$w+x+y+z$

$$P(A|B) \times P(B) = \frac{w}{w+y} \times \frac{w+y}{w+x+y+z} = \frac{w}{w+x+y+z}$$

$$P(B|A) \times P(A) = \frac{w}{w+x} \times \frac{w+x}{w+x+y+z} = \frac{w}{w+x+y+z}$$

A geometric visualisation of Bayes' theorem. In the table, the values w, x, y and z give the relative weights of each corresponding condition and case. The figures denote the cells of the table involved in each metric, the probability being the fraction of each figure that is shaded. This shows that $P(A|B)\,P(B) = P(B|A)\,P(A)$ i.e. $P(A|B) = P(B|A)\,P(A)/P(B)$. Similar reasoning can be used to show that $P(\bar{A}|B) = P(B|\bar{A})\,P(\bar{A})/P(B)$ etc.

In general, it cannot be assumed that $P(A|B) \approx P(B|A)$. This can be an insidious error, even for those who are highly conversant with statistics. The relationship between $P(A|B)$ and $P(B|A)$ is given by Bayes' theorem:

$$\begin{aligned} P(B \mid A) &= \frac{P(A \mid B)P(B)}{P(A)} \\ \Leftrightarrow \frac{P(B \mid A)}{P(A \mid B)} &= \frac{P(B)}{P(A)} \end{aligned}$$

That is, $P(A|B) \approx P(B|A)$ only if $P(B)/P(A) \approx 1$, or equivalently, $P(A) \approx P(B)$.

Assuming Marginal and Conditional Probabilities are of Similar Size

In general, it cannot be assumed that $P(A) \approx P(A|B)$. These probabilities are linked through the law of total probability:

$$P(A) = \sum_n P(A \cap B_n) = \sum_n P(A \mid B_n)P(B_n).$$

where the events (B_n) form a countable partition of A.

This fallacy may arise through selection bias. For example, in the context of a medical claim, let S_C be the event that a sequela (chronic disease) S occurs as a consequence of circumstance (acute condition) C. Let H be the event that an individual seeks medical help. Suppose that in most cases, C does not cause S so $P(S_C)$ is low. Suppose also that medical attention is only sought if S has occurred due to C. From experience of patients, a doctor may therefore erroneously conclude that $P(S_C)$ is high. The actual probability observed by the doctor is $P(S_C|H)$.

Over- or Under-weighting Priors

Not taking prior probability into account partially or completely is called *base rate neglect*. The reverse, insufficient adjustment from the prior probability is *conservatism*.

Formal Derivation

Formally, $P(A|B)$ is defined as the probability of A according to a new probability function on the sample space, such that outcomes not in B have probability 0 and that it is consistent with all original probability measures.

Let Ω be a sample space with elementary events $\{\omega\}$. Suppose we are told the event $B \subseteq \Omega$ has occurred. A new probability distribution (denoted by the conditional notation) is to be assigned on $\{\omega\}$ to reflect this. For events in B, it is reasonable to assume that the relative magnitudes of the probabilities will be preserved. For some constant scale factor α, the new distribution will therefore satisfy:

$$1.\ \omega \in B : P(\omega \mid B) = \alpha P(\omega)$$

$$2.\ \omega \notin B : P(\omega \mid B) = 0$$

$$3.\ \sum_{\omega \in \Omega} P(\omega \mid B) = 1.$$

Substituting 1 and 2 into 3 to select α:

$$\begin{aligned} 1 &= \sum_{\omega \in \Omega} P(\omega \mid B) \\ &= \sum_{\omega \in B} \alpha P(\omega) + \sum_{\omega \notin B} 0^{0} \\ &= \alpha \sum_{\omega \in B} P(\omega) \\ &= \alpha \cdot P(B) \\ \Rightarrow \alpha &= \frac{1}{P(B)} \end{aligned}$$

So the new probability distribution is

$$\begin{aligned} &1.\ \omega \in B \quad : P(\omega \mid B) = \frac{P(\omega)}{P(B)} \\ &2.\ \omega \notin B \quad : P(\omega \mid B) = 0 \end{aligned}$$

Now for a general event *A*,

$$\begin{aligned} P(A \mid B) &= \sum_{\omega \in A \cap B} P(\omega \mid B) + \sum_{\omega \in A \cap B^{c}} P(\omega \mid B)^{0} \\ &= \sum_{\omega \in A \cap B} \frac{P(\omega)}{P(B)} \\ &= \frac{P(A \cap B)}{P(B)} \end{aligned}$$

Many events associated to the real experiments, specifically related to civil engineering, have some dependency on the other events associated to the same experiment. In that case the occurrence of the dependent event is a function of the occurrence of the other events. One such example is given the probability of having cube density in between two values what is the probability of having the cube strength ranging in between two specific values.

The probability of the occurrence of an event A given that an event B has already occurred is called the conditional probability of A given B

$$P(A/B) = \frac{P(A \cap B)}{P(B)} \quad \text{If } P(B) \neq 0$$

The same is explained in the figure below using the sample spaces related to the events A and B, assuming that there are few sample points common to these two events. Part 1 of the figure shows the total sample space related to the experiment as in the form of rectangle and the sample space related to the event A as a circle. Similarly part 2 of the figure shows the total sample space and the sample space related to event B. As explained earlier in conditional probability the total sample space is restrained to the sample space that is related to event B (which has already occurred). The same is shown in part 3 of the figure given below. Now the sample space for event A (B is the total sample space available) is nothing but the sample points related to event A and falling in the sample space. This is nothing but the intersection of the events A and B and is shown in part 3 of the figure as the hatched area.

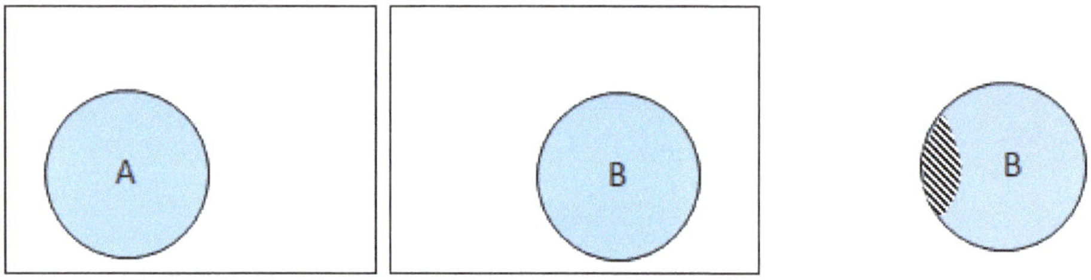

Representation of conditional probability using the Venn diagrams

For example, there are 100 trips per day between two places X and Y. Out of these 100 trips 50 are made by car, 25 are made by bus and the other 25 are by local train. Probabilities associated to these modes are 0.5, 0.25, and 0.25, respectively. In transportation engineering both the bus and the local train are considered as public transport so the event space associated to this is the summation of the event spaces associated to bus and local train. Probability of choosing public transportation is 0.5. Now if one is interested in finding the probability of choosing bus given public transportation is chosen the conditional probability is useful in finding that.

$$P(Bus/PT) = \frac{P(Bus \cap PT)}{P(PT)} = 0.5$$

Stochastic Independence

Any two events related to an experiment are stochastically independent if the probability of occurrence of the either of the two events is not influenced by the occurrence of the other event. In other words there are no common elements in the sample spaces corresponding to these two events. Two events A1 and A2 are stochastically independent if,

$P(A1/A2) = P(A1)$, for $P(A2) > 0$;
or
$P(A2/A1) = P(A2)$, for $P(A1) > 0$.

The conditional probability can be rewritten as follows if the probability of choosing bus and public transport are stochastically independent,

$$\begin{aligned} P(Bus \cap PT) &= P(Bus/PT) * P(t) \\ &= P(Bus) * P(t) \end{aligned}$$

Total Probability and Bayes' Theorem

If $A_1, A_2, A_3 \ldots. A_N$, are mutually exclusive subsets of sample space Ω then these subsets are called as partitions of Ω. If B is an arbitrary subset of Ω (shown as a circle in the following figure), then the total probability of B is given by

$$P(B) = P(A_1) . P(B/A_1) + P(A_2) . P(B/A_2) + \ldots + P(A_k) . P(B/A_k)$$

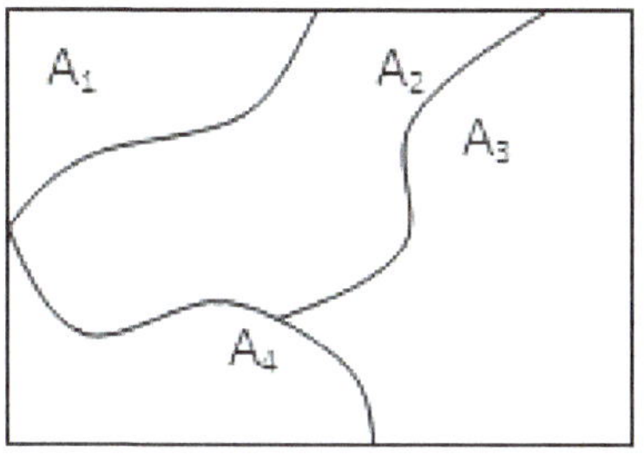

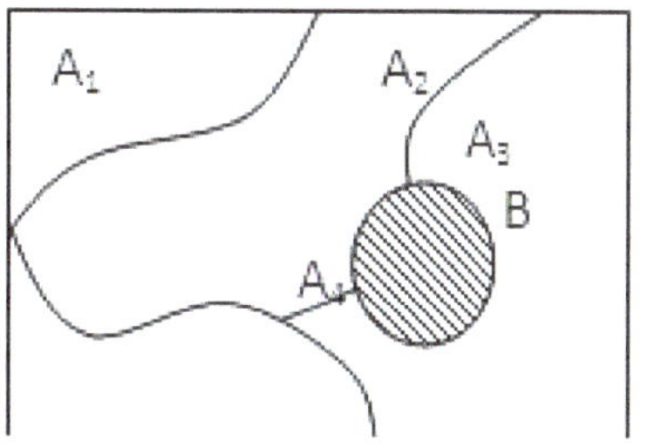

Universal set divided into various subsets

If B is an arbitrary subset of Ω, then for r = 1, 2,..., k

$$P\left(\frac{A_r}{B}\right) = \frac{P(A_r).P\left(\frac{B}{A_r}\right)}{\sum_{i=1}^{k} P(A_i) * P\left(\frac{B}{A_i}\right)}$$

Concepts from the Theory of Sets

Concepts from the theory of sets are useful in presenting the basic concepts of probability. Events in the sample space can be related in several ways. Some events may be mutually exclusive and some events may have some overlap in the sample space and it is useful to explain these interrelations among the events using the concepts of sets.

The *universal set* is the set of all the objects under consideration and it is generally denoted by letter Ω and all the subsets are denoted with the capital letters. The *Universal set* is equivalent to the sample space related to a random experiment. To represent sets, the notations used would be continued (e.g. universal set is represented using Ω). Universal set related to the 'throwing coin once' experiment is Ω = {H,T}.

Sample space related to any event is a subset of the Universal set related to that experiment. For example, in the 'tossing the coin twice' experiment the sample space corresponding to the event 'getting heads in the first toss' is A_1 = {HH, HT} and it is the subset of the universal set Ω = {HH, HT, TH, TT} corresponding to all the events resulting from the experiment. This is shown as $A_1 \subset \Omega$.

Operations on Sets

Operations on the sets related to various events can be represented using venn diagrams. The universal set is represented using a rectangle and different events, called as subsets, are shown using circles.

The *union* of two events represent the elements that occur in either of the events, and it comprises the elements that are there in either one of the sample spaces and elements in both of them. Union of two events A and B is shown in the following figure using venn diagram. Union of two events is shown using the notation AUB.

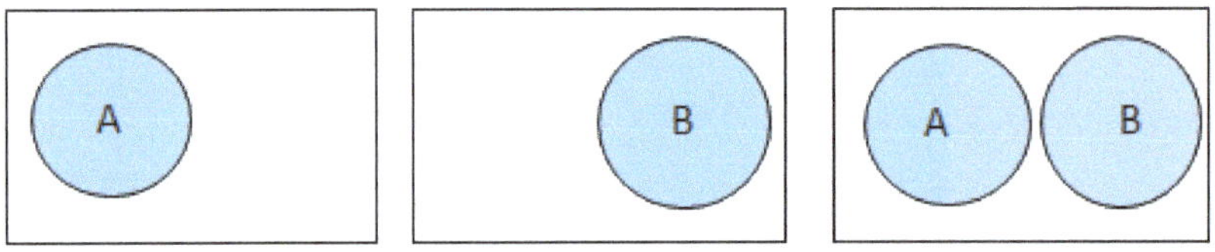

(a)

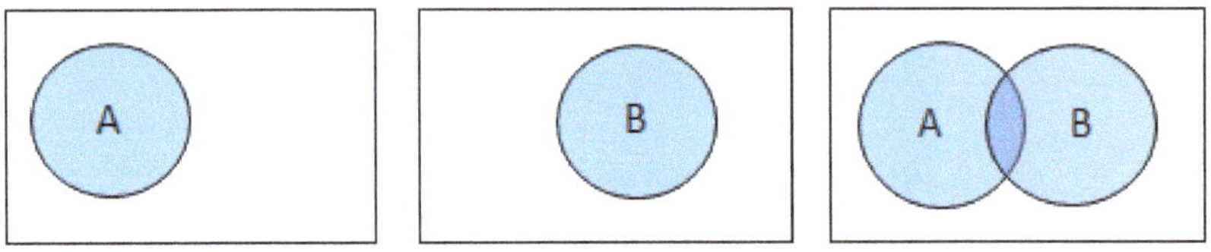

Union of two sets

The following figure (a) shows the union of events A and B which are mutually excusive. Part 1 of this figure shows the total sample space and the sample space related to the event A. Part 2 of this figure shows the sample space related the event B in the total sample space. Part 3 of the figure shows the events A and B along with the total sample space. From this part it can be seen that there is no overlap in the sample spaces related to these two events hence these two events are mutually exclsuive. In this case the union operation results in the summation of the sample spaces related to the two events.

The following figure (b) shows the union of two events A and B whose sample spaces have some overlap. 1st and 2nd parts of Figure (b) shows the sample spaces related to the two events A and B and the 3rd part shows both the events. It can be seen from this figure that some sample points are common to both the events. When union operation is carried out on these events, simple summation results in the repetition of the common sample points. To avoid this, from the summation the common points have to be deducted.

The intersection of two events represent the common occurrence, and it comprises the elements that are common in both the sample spaces. Intersection of two events A and B is shown in the following figure using venn diagram. Intersection of two events is shown as A ∩ B or AB.

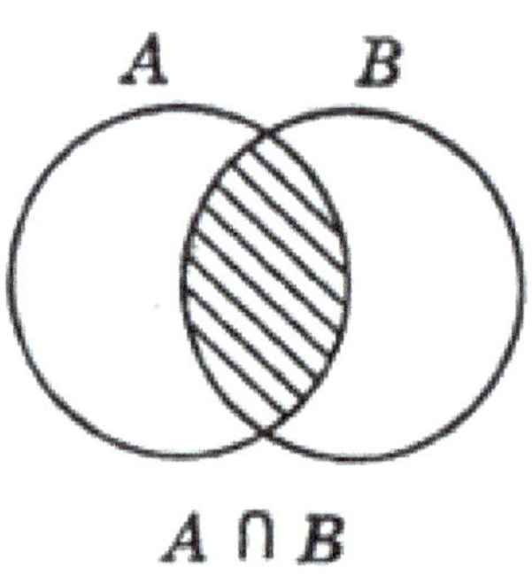

Intersection of two sets

If the occurrence of one event prevents the occurrence of the other event then they are called mutually exclusive or disjoint. For example, two events associated to having water levels in the supply storage level and in the flood control storage level are mutually exclusive. If event A represents the water level in the supply storage level and the event B represents the water level in the flood control storage level, together A and B in the form A∩Bconstitute a null event denoted by Φ. If the two events are not mutually exclusive the intersection represents the common sample space.

The operations on sets can be directly extended to more than two events. One such operation is shown in the figure below, where the operation A ∩ (BUC) is shown.

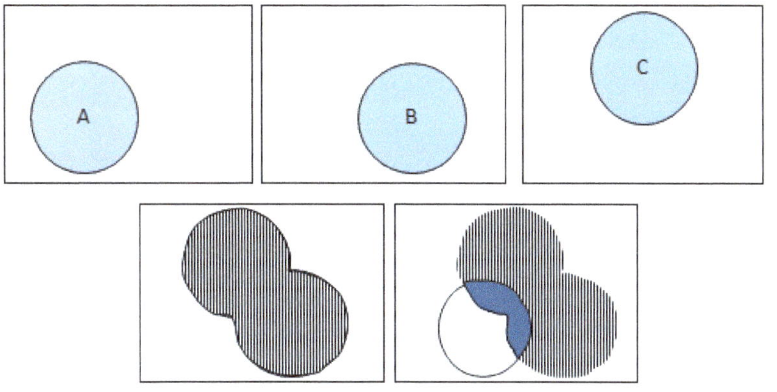

A∩(BUC)

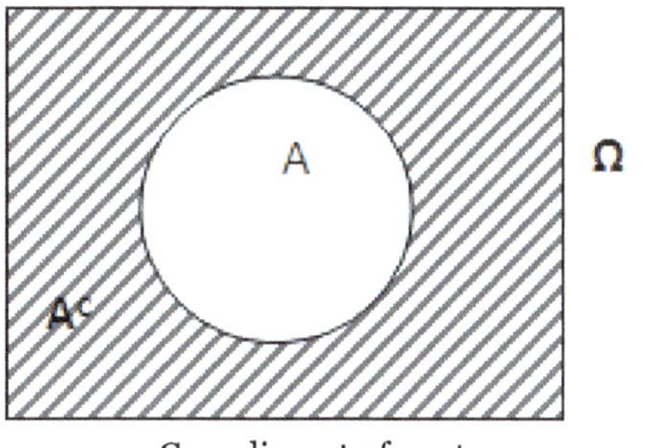

Compliment of a set

If A is the subset of the universal set, Ω, the hatched portion is complement of A denoted with A^c.

Different laws such as cumulative law, associative law, and distributive law, which are used in algebra, are also applicable to the set operations. For example, associative law between three events (subsets) A, B, and C is (AUB) UC = AU(BUC) which is similar to the one used in algebra. The same can be seen in the figure below.

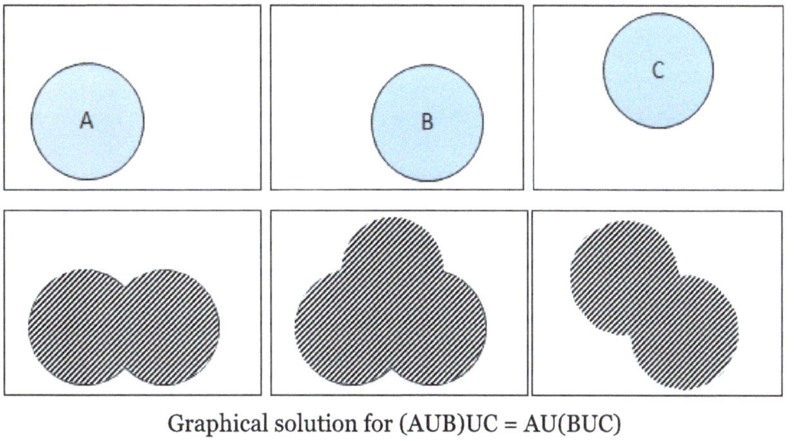

Graphical solution for (AUB)UC = AU(BUC)

Parts 1, 2, and 3 in the figure given above are showing the sample spaces related to the three events A, B, and C, respectively. Part 4 shows the union operation on events A and B. Part 6 shows the union operation on events B and C. Part 5 results from either the union operation on the sample space related to the union of A and B (Part 4) and the sample space related to event C or the union operation on the sample space related to the union of B and C (part 6) and the sample space related to event A.

3

Random Variable: An Overview

Random variables measure the probability of a random phenomenon. While the process requires dealing with real numbers and accurate probabilities, the logic behind constituting the phenomenon itself is random or not entirely known. The aspects elucidated in this chapter are of vital importance, and provide a better understanding of random variables.

Random Variable

In probability and statistics, a random variable, random quantity, aleatory variable, or stochastic variable is a variable quantity whose value depends on possible outcomes. As a function, a random variable is required to be measurable, which rules out certain pathological cases where the quantity which the random variable returns is infinitely sensitive to small changes in the outcome.

It is common that these outcomes depend on some physical variables that are not well understood. For example, when you toss a coin, the final outcome of heads or tails depends on the uncertain physics. Which outcome will be observed is not certain. Of course the coin could get caught in a crack in the floor, but such a possibility is excluded from consideration. The domain of a random variable is the set of possible outcomes. In the case of the coin, there are only two possible outcomes, namely heads or tails. Since one of these outcomes must occur, thus either the event that the coin lands heads or the event that the coin lands tails must have non-zero probability.

A random variable is defined as a function that maps outcomes to numerical quantities (labels), typically real numbers. In this sense, it is a procedure for assigning a numerical quantity to each physical outcome, and, contrary to its name, this procedure itself is neither random nor variable. What is random is the unstable physics that describes how the coin lands, and the uncertainty of which outcome will actually be observed.

A random variable's possible values might represent the possible outcomes of a yet-to-be-performed experiment, or the possible outcomes of a past experiment whose already-existing value is uncertain (for example, due to imprecise measurements or quantum uncertainty). They may also conceptually represent either the results of an "objectively" random process (such as rolling a die) or the "subjective" randomness that results from incomplete knowledge of a quantity. The meaning of the probabilities assigned to the potential values of a random variable is not part of probability theory itself but is instead related to philosophical arguments over the interpretation of probability. The mathematics works the same regardless of the particular interpretation in use.

A random variable has a probability distribution, which specifies the probability that its value falls in any given interval. Random variables can be discrete, that is, taking any of a specified finite or countable list of values, endowed with a probability mass function characteristic of the random variable's probability distribution; or continuous, taking any numerical value in an interval or collection of intervals, via a probability density function that is characteristic of the random variable's probability distribution; or a mixture of both types. Two random variables with the same probability distribution can still differ in terms of their associations with, or independence from, other random variables. The realizations of a random variable, that is, the results of randomly choosing values according to the variable's probability distribution function, are called random variates.

The formal mathematical treatment of random variables is a topic in probability theory. In that context, a random variable is understood as a function defined on a sample space whose outputs are numerical values.

Definition

A *random variable* $X : \Omega \to E$ is a measurable function from a set of possible outcomes Ω to a measurable space E. The technical axiomatic definition requires Ω to be a probability space. Usually X is real-valued (i.e. $E = \mathbb{R}$).

A random variable does *not* return a probability. The probability of a measurable set of outcomes (i.e. an event) is given by the probability measure P with which Ω is equipped. Rather, X returns a numerical quantity of outcomes in Ω — e.g. the number of heads in a random collection of coin flips, or the height of a random person. The probability that X takes value ≤ 3 is the probability of the set of outcomes $\{\omega \in \Omega : X(\omega) \leq 3\}$, denoted $P(X \leq 3)$.

Standard Case

In many cases, $E = \mathbb{R}$. In some contexts, the term *random element* is used to denote a random variable not of this form.

When the image (or range) of X is finite or countably infinite, the random variable is called a discrete random variable and its distribution can be described by a probability mass function which assigns a probability to each value in the image of X. If the image is uncountably infinite then X is called a continuous random variable. In the special case that it is absolutely continuous, its distribution can be described by a probability density function, which assigns probabilities to intervals; in particular, each individual point must necessarily have probability zero for an absolutely continuous random variable. Not all continuous random variables are absolutely continuous, for example a mixture distribution. Such random variables cannot be described by a probability density or a probability mass function.

Any random variable can be described by its cumulative distribution function, which describes the probability that the random variable will be less than or equal to a certain value.

Extensions

The term "random variable" in statistics is traditionally limited to the real-valued case ( ). In this case, the structure of the real numbers makes it possible to define quantities such as the expected value and variance of a random variable, its cumulative distribution function, and the moments of its distribution.

However, the definition above is valid for any measurable space E of values. Thus one can consider random elements of other sets E, such as random boolean values, categorical values, complex numbers, vectors, matrices, sequences, trees, sets, shapes, manifolds, and functions. One may then specifically refer to a *random variable of type* E, or an E*-valued random variable.*

This more general concept of a random element is particularly useful in disciplines such as graph theory, machine learning, natural language processing, and other fields in discrete mathematics and computer science, where one is often interested in modeling the random variation of non-numerical data structures. In some cases, it is nonetheless convenient to represent each element of E using one or more real numbers. In this case, a random element may optionally be represented as a vector of real-valued random variables (all defined on the same underlying probability space Ω, which allows the different random variables to covary). For example:

- A random word may be represented as a random integer that serves as an index into the vocabulary of possible words. Alternatively, it can be represented as a random indicator vector whose length equals the size of the vocabulary, where the only values of positive probability are $(1000\cdots)$, $(0100\cdots)$, $(0010\cdots)$ and the position of the 1 indicates the word.
- A random sentence of given length N may be represented as a vector of N random words.
- A random graph on N given vertices may be represented as a $N\times N$ matrix of random variables, whose values specify the adjacency matrix of the random graph.
- A random function F may be represented as a collection of random variables $F(x)$, giving the function's values at the various points x in the function's domain. The $F(x)$ are ordinary real-valued random variables provided that the function is real-valued. For example, a stochastic process is a random function of time, a random vector is a random function of some index set such as $1,2,\ldots n$, and random field is a random function on any set (typically time, space, or a discrete set).

Examples

Discrete Random Variable

In an experiment a person may be chosen at random, and one random variable may be the person's height. Mathematically, the random variable is interpreted as a function which maps the person to the person's height. Associated with the random variable is a probability distribution that allows the computation of the probability that the height is in any subset of possible values, such as the probability that the height is between 180 and 190 cm, or the probability that the height is either less than 150 or more than 200 cm.

Another random variable may be the person's number of children; this is a discrete random variable with non-negative integer values. It allows the computation of probabilities for individual integer values – the probability mass function (PMF) – or for sets of values, including infinite sets. For example, the event of interest may be "an even number of children". For both finite and infinite event sets, their probabilities can be found by adding up the PMFs of the elements; that is, the probability of an even number of children is the infinite sum $\mathrm{PMF}(0)+\mathrm{PMF}(2)+\mathrm{PMF}(4)+\cdots$.

In examples such as these, the sample space (the set of all possible persons) is often suppressed, since it is mathematically hard to describe, and the possible values of the random variables are then treated as a sample space. But when two random variables are measured on the same sample space of outcomes, such as the height and number of children being computed on the same random persons, it is easier to track their relationship if it is acknowledged that both height and number of children come from the same random person, for example so that questions of whether such random variables are correlated or not can be posed.

Coin Toss

The possible outcomes for one coin toss can be described by the sample space $\Omega = \{\text{heads}, \text{tails}\}$. We can introduce a real-valued random variable Y that models a \$1 payoff for a successful bet on heads as follows:

$$Y(\omega) = \begin{cases} 1, & \text{if } \omega = \text{heads}, \\ 0, & \text{if } \omega = \text{tails}. \end{cases}$$

If the coin is a fair coin, Y has a probability mass function f_Y given by:

$$f_Y(y) = \begin{cases} \frac{1}{2}, & \text{if } y = 1, \\ \frac{1}{2}, & \text{if } y = 0, \end{cases}$$

Dice Roll

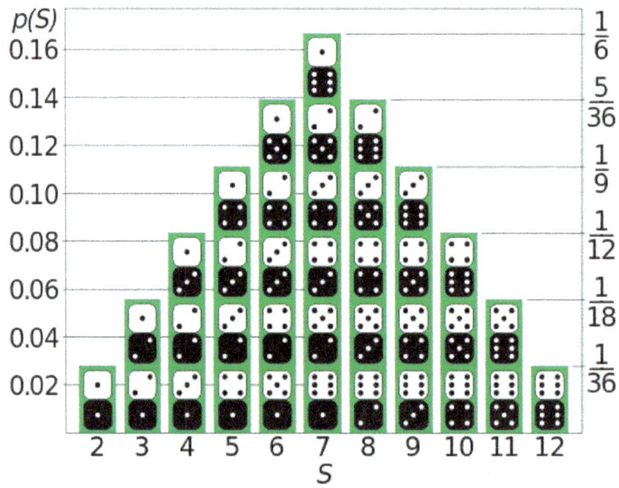

If the sample space is the set of possible numbers rolled on two dice, and the random variable of interest is the sum S of the numbers on the two dice, then S is a discrete random variable whose distribution is described by the probability mass function plotted as the height of picture columns here.

A random variable can also be used to describe the process of rolling dice and the possible outcomes. The most obvious representation for the two-dice case is to take the set of pairs of numbers

n_1 and n_2 from {1, 2, 3, 4, 5, 6} (representing the numbers on the two dice) as the sample space. The total number rolled (the sum of the numbers in each pair) is then a random variable X given by the function that maps the pair to the sum:

$$X((n_1, n_2)) = n_1 + n_2$$

and (if the dice are fair) has a probability mass function f_X given by:

$$f_X(S) = \tfrac{\min(S-1,13-S)}{36}, \text{for } S \in \{2,3,4,5,6,7,8,9,10,11,12\}$$

Continuous Random Variable

An example of a continuous random variable would be one based on a spinner that can choose a horizontal direction. Then the values taken by the random variable are directions. We could represent these directions by North, West, East, South, Southeast, etc. However, it is commonly more convenient to map the sample space to a random variable which takes values which are real numbers. This can be done, for example, by mapping a direction to a bearing in degrees clockwise from North. The random variable then takes values which are real numbers from the interval [0, 360), with all parts of the range being "equally likely". In this case, X = the angle spun. Any real number has probability zero of being selected, but a positive probability can be assigned to any *range* of values. For example, the probability of choosing a number in [0, 180] is ½. Instead of speaking of a probability mass function, we say that the probability *density* of X is 1/360. The probability of a subset of [0, 360) can be calculated by multiplying the measure of the set by 1/360. In general, the probability of a set for a given continuous random variable can be calculated by integrating the density over the given set.

Mixed Type

An example of a random variable of mixed type would be based on an experiment where a coin is flipped and the spinner is spun only if the result of the coin toss is heads. If the result is tails, $X = -1$; otherwise X = the value of the spinner as in the preceding example. There is a probability of ½ that this random variable will have the value −1. Other ranges of values would have half the probabilities of the last example.

Measure-theoretic Definition

The most formal, axiomatic definition of a random variable involves measure theory. Continuous random variables are defined in terms of sets of numbers, along with functions that map such sets to probabilities. Because of various difficulties (e.g. the Banach–Tarski paradox) that arise if such sets are insufficiently constrained, it is necessary to introduce what is termed a sigma-algebra to constrain the possible sets over which probabilities can be defined. Normally, a particular such sigma-algebra is used, the Borel σ-algebra, which allows for probabilities to be defined over any sets that can be derived either directly from continuous intervals of numbers or by a finite or countably infinite number of unions and/or intersections of such intervals.

The measure-theoretic definition is as follows.

Let $(\Omega, \mathcal{F}, P)$ be a probability space and $(E, \mathcal{E})$ a measurable space. Then an $(E, \mathcal{E})$-valued random variable is a function $X : \Omega \to E$ which is $(\mathcal{F}, \mathcal{E})$-measurable. The latter means that, for every subset $B \in \mathcal{E}$, its preimage $X^{-1}(B) \in \mathcal{F}$ where $X^{-1}(B) = \{\omega : X(\omega) \in B\}$. This definition enables us to measure any subset $B \in \mathcal{E}$ in the target space by looking at its preimage, which by assumption is measurable.

In more intuitive terms, a member of Ω is a possible outcome, a member of $\mathcal{F}$ is a measurable subset of possible outcomes, the function P gives the probability of each such measurable subset, E represents the set of values that the random variable can take (such as the set of real numbers), and a member of $\mathcal{E}$ is a "well-behaved" (measurable) subset of E (those for which you might want to find the probability). The random variable is then a function from any outcome to a quantity, such that the outcomes leading to any useful subset of quantities for the random variable have a well-defined probability.

When E is a topological space, then the most common choice for the σ-algebra $\mathcal{E}$ is the Borel σ-algebra $\mathcal{B}(E)$, which is the σ-algebra generated by the collection of all open sets in E. In such case the $(E, \mathcal{E})$-valued random variable is called the E-valued random variable. Moreover, when space E is the real line $\mathbb{R}$, then such a real-valued random variable is called simply the random variable.

Real-valued Random Variables

In this case the observation space is the set of real numbers. Recall, $(\Omega, \mathcal{F}, P)$ is the probability space. For real observation space, the function $X : \Omega \to \mathbb{R}$ is a real-valued random variable if

$$\{\omega : X(\omega) \le r\} \in \mathcal{F} \qquad \forall r \in \mathbb{R}.$$

This definition is a special case of the above because the set $\{(-\infty, r] : r \in \mathbb{R}\}$ generates the Borel σ-algebra on the set of real numbers, and it suffices to check measurability on any generating set. Here we can prove measurability on this generating set by using the fact that $\{\omega : X(\omega) \le r\} = X^{-1}((-\infty, r])$.

Distribution Functions of Random Variables

If a random variable $X : \Omega \to \mathbb{R}$ defined on the probability space $(\Omega, \mathcal{F}, P)$ is given, we can ask questions like "How likely is it that the value of X is equal to 2?". This is the same as the probability of the event $\{\omega : X(\omega) = 2\}$ which is often written as $P(X = 2)$ or $p_X(2)$ for short.

Recording all these probabilities of output ranges of a real-valued random variable X yields the probability distribution of X. The probability distribution "forgets" about the particular probability space used to define X and only records the probabilities of various values of X. Such a probability distribution can always be captured by its cumulative distribution function

$$F_X(x) = \mathrm{P}(X \le x)$$

and sometimes also using a probability density function, p_X. In measure-theoretic terms, we use the random variable X to "push-forward" the measure P on Ω to a measure p_X on $\mathbb{R}$. The underlying probability space Ω is a technical device used to guarantee the existence of random variables, sometimes to construct them, and to define notions such as correlation and dependence or independence based on a joint distribution of two or more random variables on the same probability space. In practice, one often disposes of the space Ω altogether and just puts a measure on $\mathbb{R}$ that assigns measure 1 to the whole real line, i.e., one works with probability distributions instead of random variables.

Moments

The probability distribution of a random variable is often characterised by a small number of parameters, which also have a practical interpretation. For example, it is often enough to know what its "average value" is. This is captured by the mathematical concept of expected value of a random variable, denoted $\mathrm{E}[X]$, and also called the first moment. In general, $\mathrm{E}[f(X)]$ is not equal to $f(\mathrm{E}[X])$. Once the "average value" is known, one could then ask how far from this average value the values of X typically are, a question that is answered by the variance and standard deviation of a random variable. $\mathrm{E}[X]$ can be viewed intuitively as an average obtained from an infinite population, the members of which are particular evaluations of X.

Mathematically, this is known as the (generalised) problem of moments: for a given class of random variables X, find a collection $\{f_i\}$ of functions such that the expectation values $\mathrm{E}[f_i(X)]$ fully characterise the distribution of the random variable X.

Moments can only be defined for real-valued functions of random variables (or complex-valued, etc.). If the random variable is itself real-valued, then moments of the variable itself can be taken, which are equivalent to moments of the identity function $f(X) = X$ of the random variable. However, even for non-real-valued random variables, moments can be taken of real-valued functions of those variables. For example, for a categorical random variable X that can take on the nominal values "red", "blue" or "green", the real-valued function $[X = \text{green}]$ can be constructed; this uses the Iverson bracket, and has the value 1 if X has the value "green", 0 otherwise. Then, the expected value and other moments of this function can be determined.

Functions of Random Variables

A new random variable Y can be defined by applying a real Borel measurable function $g:\mathbb{R}\rightarrow\mathbb{R}$ to the outcomes of a real-valued random variable X. The cumulative distribution function of Y is

$$F_Y(y) = \mathrm{P}(g(X) \leq y).$$

If function g is invertible, i.e., g^{-1} exists, and is either increasing or decreasing, then the previous relation can be extended to obtain

$$F_Y(y) = \mathrm{P}(g(X) \leq y) = \begin{cases} \mathrm{P}(X \leq g^{-1}(y)) = F_X(g^{-1}(y)), & \text{if } g^{-1} \text{ increasing}, \\ \mathrm{P}(X \geq g^{-1}(y)) = 1 - F_X(g^{-1}(y)), & \text{if } g^{-1} \text{ decreasing}. \end{cases}$$

and, again with the same hypotheses of invertibility of g, assuming also differentiability, we can find the relation between the probability density functions by differentiating both sides with respect to y, in order to obtain

$$f_Y(y) = f_X(g^{-1}(y))\left|\frac{dg^{-1}(y)}{dy}\right|.$$

If there is no invertibility of g but each y admits at most a countable number of roots (i.e., a finite, or countably infinite, number of x_i such that $y = g(x_i)$ then the previous relation between the probability density functions can be generalized with

$$f_Y(y) = \sum_i f_X(g_i^{-1}(y))\left|\frac{dg_i^{-1}(y)}{dy}\right|$$

where $x_i = g_i^{-1}(y)$. The formulas for densities do not demand g to be increasing.

In the measure-theoretic, axiomatic approach to probability, if we have a random variable X on Ω and a Borel measurable function $g:\mathbb{R}\to\mathbb{R}$, then $Y = g(X)$ will also be a random variable on Ω, since the composition of measurable functions is also measurable. (However, this is not true if g is Lebesgue measurable.) The same procedure that allowed one to go from a probability space (Ω, P) to $(\mathbb{R}, dF_X)$ can be used to obtain the distribution of Y .

Example 1

Let X be a real-valued, continuous random variable and let $Y = X^2$.

$$F_Y(y) = \mathrm{P}(X^2 \leq y).$$

If $y < 0$, then $P(X^2 \leq y) = 0$, so

$$F_Y(y) = 0 \qquad \text{if} \quad y < 0.$$

If $y \geq 0$, then

$$\mathrm{P}(X^2 \leq y) = \mathrm{P}(|X| \leq \sqrt{y}) = \mathrm{P}(-\sqrt{y} \leq X \leq \sqrt{y}),$$

so

$$F_Y(y) = F_X(\sqrt{y}) - F_X(-\sqrt{y}) \qquad \text{if} \quad y \geq 0.$$

Example 2

Suppose X is a random variable with a cumulative distribution

$$F_X(x) = P(X \leq x) = \frac{1}{(1+e^{-x})^{\theta}}$$

where $\theta > 0$ is a fixed parameter. Consider the random variable $Y = \log(1+e^{-X})$. Then,

$$F_Y(y) = P(Y \leq y) = P(\log(1+e^{-X}) \leq y) = P(X \geq -\log(e^y - 1)).$$

The last expression can be calculated in terms of the cumulative distribution of X, so

$$F_Y(y) = 1 - F_X(-\log(e^y - 1))$$

$$= 1 - \frac{1}{(1 + e^{\log(e^y - 1)})^\theta}$$

$$= 1 - \frac{1}{(1 + e^y - 1)^\theta}$$

$$= 1 - e^{-y\theta}.$$

which is the cumulative distribution function (cdf) of an exponential distribution.

Example 3

Suppose X is a random variable with a standard normal distribution, whose density is

$$f_X(x) = \frac{1}{\sqrt{2\pi}} e^{-x^2/2}.$$

Consider the random variable $Y = X^2$. We can find the density using the above formula for a change of variables:

$$f_Y(y) = \sum_i f_X(g_i^{-1}(y)) \left| \frac{dg_i^{-1}(y)}{dy} \right|.$$

In this case the change is not monotonic, because every value of Y has two corresponding values of X (one positive and negative). However, because of symmetry, both halves will transform identically, i.e.,

$$f_Y(y) = 2 f_X(g^{-1}(y)) \left| \frac{dg^{-1}(y)}{dy} \right|.$$

The inverse transformation is

$$x = g^{-1}(y) = \sqrt{y}$$

and its derivative is

$$\frac{dg^{-1}(y)}{dy} = \frac{1}{2\sqrt{y}}.$$

Then,

$$f_Y(y) = 2\frac{1}{\sqrt{2\pi}}e^{-y/2}\frac{1}{2\sqrt{y}} = \frac{1}{\sqrt{2\pi y}}e^{-y/2}.$$

This is a chi-squared distribution with one degree of freedom.

Equivalence of Random Variables

There are several different senses in which random variables can be considered to be equivalent. Two random variables can be equal, equal almost surely, or equal in distribution.

In increasing order of strength, the precise definition of these notions of equivalence is given below.

Equality in Distribution

If the sample space is a subset of the real line, random variables X and Y are *equal in distribution* (denoted $X \overset{d}{=} Y$) if they have the same distribution functions:

$$P(X \le x) = P(Y \le x) \quad \text{for all} \quad x.$$

Two random variables having equal moment generating functions have the same distribution. This provides, for example, a useful method of checking equality of certain functions of i.i.d. random variables. However, the moment generating function exists only for distributions that have a defined Laplace transform.

Almost Sure Equality

Two random variables X and Y are *equal almost surely* if, and only if, the probability that they are different is zero:

$$P(X \ne Y) = 0.$$

For all practical purposes in probability theory, this notion of equivalence is as strong as actual equality. It is associated to the following distance:

$$d_\infty(X,Y) = \operatorname*{ess\,sup}_{\omega} |X(\omega) - Y(\omega)|,$$

where "ess sup" represents the essential supremum in the sense of measure theory.

Equality

Finally, the two random variables X and Y are *equal* if they are equal as functions on their measurable space:

$$X(\omega) = Y(\omega) \qquad \text{for all } \omega.$$

Convergence

A significant theme in mathematical statistics consists of obtaining convergence results for certain sequences of random variables; for instance the law of large numbers and the central limit theorem.

There are various senses in which a sequence (X_n) of random variables can converge to a random variable *X*.

A random variable may be defined as a variable whose values depend on the outcomes of a random experiment. In a more probabilistic sense it can be defined as follows;

'If E is an experiment having sample space Ω and X is a function that assigns a real number X(e) to every outcome e E, then X(e) is called a random variable'.

Table: Outcomes of the 'tossing the coin twice' experiment

Outcome	**HH**	**HT**	**TH**	**TT**
X	2	1	1	0

The domain of X is the sample space Ω and the numbers in the range are real numbers. The table above is showing the outcomes of 'tossing the coin twice' experiment and the associated sample space. The outcomes are not numbers hence using the random variable X(e), real numbers are attached to each of the outcomes which are also shown in the table. For random variable X(e), for every outcome e that belongs to the sample space there is exactly one value of X(e) = x, for example X(HH) = 2. As discussed earlier the range is made up of the possible values of X. In this example the outcomes are equally likely to occur and each has the probability of ¼. Different possible values for X, i.e. occurrence of different events which are the subsets of the sample space, have different associated probabilities (e.g. P(X=1) = ½).

Here onwards, the upper case alphabets are used to denote the random variables and the corresponding lower case alphabets are used to denote particular values taken by the random variables.

Discrete Random Variable

A discrete random variable takes countable number of values or the range is finite. In general a discrete random variable takes integer values but not limited to only integer values. Let X be a discrete random variable which takes the values x 1 , x 2 ... , the function f(x) defined by f(x) = P(X = x) is called the probability function or the probability mass function of X.

Let f(x) is the probability function of a discrete random variable X. then we have

$$\sum_{\forall x} f(x) = 1$$

$$f(x) \geq 0, \forall x$$

Properties of the Probability Mass Function

1. The probability of a discrete random variable X taking a specific value x is $p(x)$.

$$P(X = x) = P_X(x) = P(x)$$

2. $P(x)$ is non-negative for all real x.

$$P(X = x) \; or \; P(x) > 0 \; for \; all \; possible \; x \in R$$

3. The sum of $p(x)$ over all possible values of x is 1, that is

$$\sum_{\forall x_i} P_X(x_i) = 1$$

Where, $\boldsymbol{i}$ represent all the possible values that x can take and P_i is the probability of random variable X taking the value x_i.

From properties 2 and 3it can be said that $0 \Leftarrow P(x) \Leftarrow 1$.

Consider a random variable X defined by the number of heads obtained when two balanced or true coins are tossed, the possible values of X are 0, 1, 2. The resulting probability distribution is described in table below.

Table: probability distribution function associated to the experiment "tossing two true coins"

x	0	1	2
f(x)	1/4	1/2	1/4

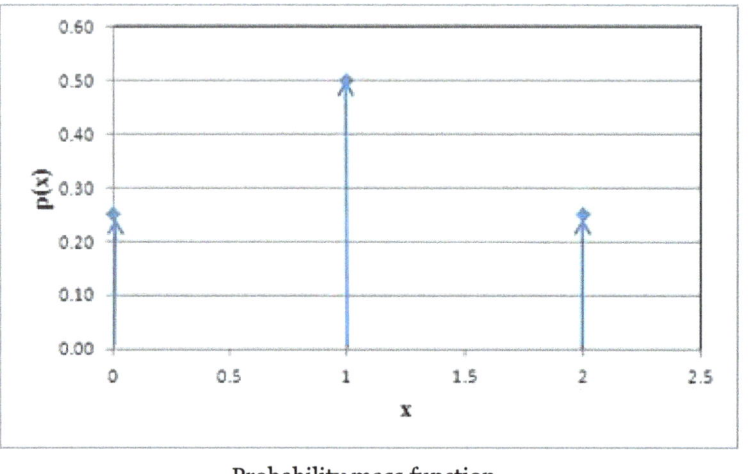

Probability mass function

Cumulative Distribution Function for a Discrete Random Variable

Cumulative distribution function (*cdf*) of any random variable is denoted by $F_X(x)$. This is also denoted by the probability of the random variable X taking the values that are smaller than or equals to x, $P(X \leq x)$. This is a monotonically increasing function bounded in between 0 and 1.

For discrete random variable X,

$$F_X(x) = \sum_{X \leq x} P_X(x)$$

Cumulative distribution function for the data on experiment "tossing the coin twice".

Another example for the discrete random variables is the number of vehicles crossing a particular location of the road stretch in a given time interval. If the time interval is 1 minute and the road is a single lane road, the possible integer values taken by the discrete random variable may fall in between 1 to 30. Sometimes the number of vehicles arrived in consecutive time intervals may not depend on each other and sometimes may depend on one another. Another example is on a particular day whether it rains or not.

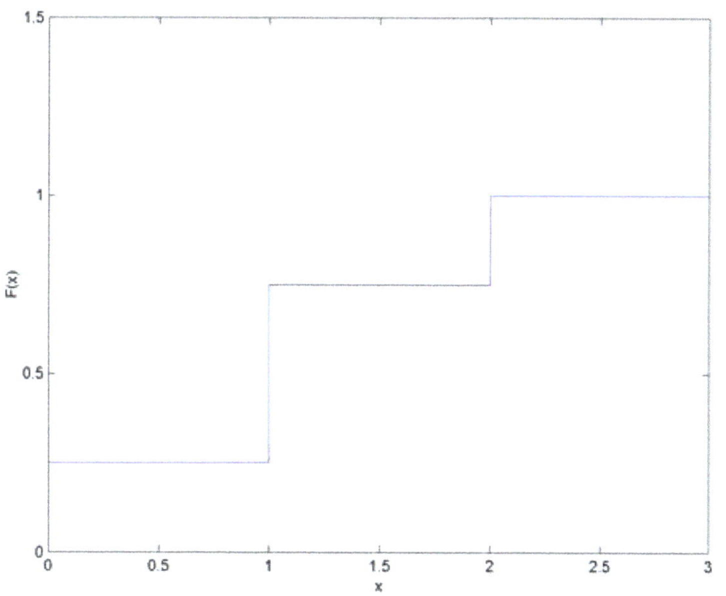

Cumulative distribution function of a discrete random variable

Continuous Random Variable

Probability Density Function

Let X is a continuous random variable. A continuous random variable takes any value between two limits, and in most of the cases it is assumed that it takes any value between -∞ and +∞. But in reality most of the continuous random variables such as compressive strength, unit weight of the soil sample etc., takes any real number between two limits.

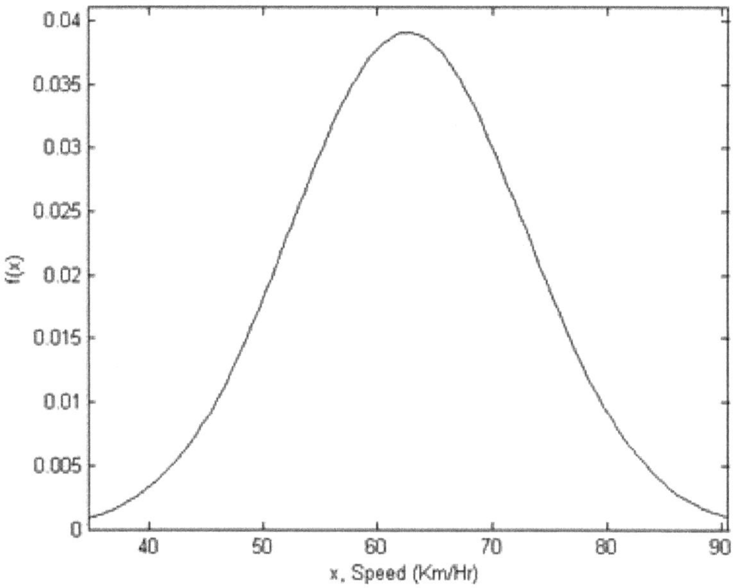

Probability density function of a continuous random variable

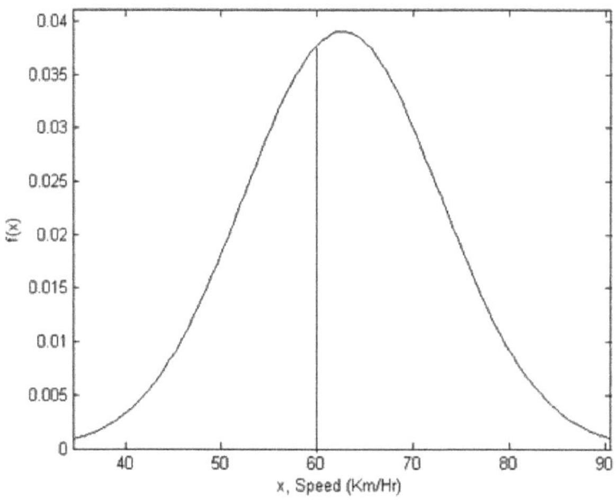

Probability density function of a continuous random variable

Associated with X there is a real valued function *f(x)* such that, for any real numbers a and b, we have,

$$p(a < X < b) = \int_a^b f(x)\,dx$$

This function is called as probability density function. Its graphical representation takes a continuous form of a curve.

For any real number a, we have

$$P(X = a) = P(a \le X \le a)$$
$$= \int_a^a f(x)\,dx = 0$$

From this it can be said that the function itself doesn't represent the probability. In other words, probability of a random variable taking a particular value (exactly that value) is zero. For example, probability of the headway between two consecutive vehicular arrivals equals to two seconds may not be possible in reality. If the headway measuring error tends to zero the measured headway may be 2±0.00001 seconds.

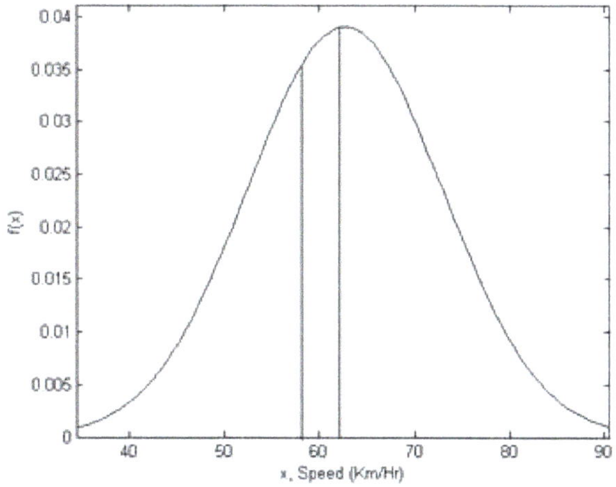

Probability density function of a continuous random variable

As described below the area under this curve gives the probability associated the random variable that takes the values between the upper and lower limits of the area. This is shown in the figure above where the area under the curve and between the two vertical lines gives the probability of having the vehicle's speed between 58.5 and 61.5 km/hr.

For any real numbers a and b, we have

$$P(\mathrm{a} \leq \mathrm{X} \leq \mathrm{b}) P(\mathrm{a}<\mathrm{X} \leq \mathrm{b}) == (\mathrm{a}<\mathrm{X}<\mathrm{b}) = P(\mathrm{a} \leq \mathrm{X}<\mathrm{b})$$

The total area under the curve is equals to one. If the random variable takes any value between -∞ and +∞ then the area under the curve is one, as shown below.

$$\int_{-\infty}^{\infty} f(x)\,dx = 1$$

The CDF for a Continuous Random Variable

The distribution function also called as the cumulative distribution function of a random variable X is the function F(x) defined by,

$$F(x) P(X \leq x) = \int_{-\infty}^{x} f(x)\,dx$$

This function gives the probability of the random variable X taking any value less than or equals to x. The pdf can be obtained by differentiating the cdf (as this is always continuous), as shown below. CDF for the vehicular speed data observed at a road section is shown in the figure below.

$$\frac{dF_x(x)}{dx} = f(x)$$

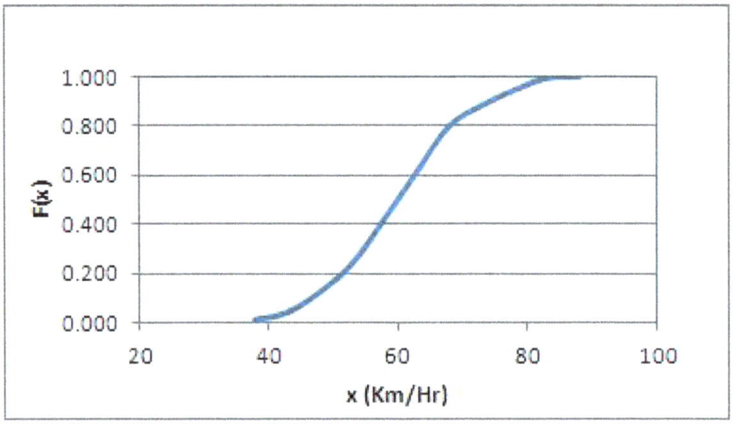

Cumulative distribution function of a continuous random variable

Important properties of the cumulative distribution function are as follows;

1. F(x) takes values in between 0 and 1. As

$$0 \leq F(x) \leq 1$$

2. As the value of x tends to infinitely large negative value F(x) tend to zero.

$$\lim_{x \to -\infty} F(x) = 0$$

As the value of x tends to infinitely large positive value F(x) tend to one.

$$\lim_{x \to \infty} F(x) = 1$$

3. F_X (x) is a monotonic, non decreasing function.

F(a) ≤ F(b) for a < b

This directly results in a non-negative probability density function for the random variable X.

Example:

The vehicular arrivals (time gap between arrivals, in seconds) at a particular location of the road follow the probability density function given below.

$$f(x) = 0 \; for \; x < 0$$
$$= 0.4 * e^{-0.4x} \; for \; 0 \leq x < \infty$$

Find the following:

a. The probability of time gap between vehicle arrivals at that location is less than 5 sec.

$$p(x<5) \Rightarrow F(5) = \int_0^5 0.4 * e^{-0.4x} ds = 0.864$$

b. The probability of time gap between vehicle arrivals at that location is in between 2 and 5 seconds.

$$p(2 \leq x \leq 5) \Rightarrow F(5) - F(2) = \int_2^5 0.4 * e^{-0.4x} dx = 0.314$$

Characteristics of Random Variables

As can be seen from the previous discussion, outcome from various experiments results in different types of random variables. To understand the nature of the outcomes (values assigned to the random variable) from various experiments and to correlate them to the theoretical probability density or probability mass functions it is necessary to understand the characteristics of the random variables.

Similar to the numerical descriptors used in describing the data, random variables can also be understood using some parameters. Moment about the origin and moment about the mean gives important descriptors of the random variable. Moment about the origin results in mean and the second moment about the mean results in variance.

Moment about the Origin

Moments about the origin are called the origin moments. μ_k denotes the kth moment about the origin.

$\mu_k = \sum_i x_i^k Px(x_i)$ is the the kth moment about the origin for the discrete random variable.

$\mu_k = \int_{-\infty}^{\infty} x^k f_x(x) dx$ is the kth moment about the origin for the continuous random variable.

Moments about the Mean

Moments about mean are called central moments and μ_k denotes the kth moment about the mean.

$\mu_k = \sum_i (x_i - \mu)^k Px(x_i)$ is the kth moment about the mean for the discrete random variable.

$\mu_k = \int_{-\infty}^{\infty} (x_i - \mu)^k f_x(x) dx$ is the kth moment about the mean for the continuous random variable.

Expectations

For any random variable X and a function of random variable X, denoted with Y= H(X) (Y is also a random variable) the expected values are denoted with E(X) and E(H(X)), respectively.

If X is a discrete random variable and p(x) is the probability function of X then

$$E(H(X)) = \sum_{\forall X} H(x)p(x_i)$$

if X is a continuous random variable and f(x) is the probability density function of X then

$$E(H(X)) = \int_{-\infty}^{\infty} H(x)f(x)dx$$

The expectation E(X) is called the arithmetic mean or simply the mean of X, and is denoted by μ. The expectation $E((X-\mu)^2)$, is the variance. Origin and central moments, discussed earlier, also be expressed using the expected value operator.

Origin moments are expressed as follows;

$$\mu_k = E\left(X^K\right)$$

Central moments are expressed as follows;

$$\mu_k = E\left(\left(X - E(X)\right)^k\right)$$

When a random variable X is multiplied by a constant the resulting mean and variances are as follows;

$E(aX) = a\,E(X)$

$V(aX) = a^2\,V(X)$

Where, V(X) is the variance of the random variable X.

Expectation of a constant is the constant itself.

When the functions of the random variables are complicated then the evaluation of the mean and the variance is difficult. In that case the function may be expanded using Taylor series expansion and the expectations are applied, may be on the first few terms of the expansion, to know the mean and variance. The expansion of the complicated function is carried about the mean of the random variable X. H(X) is expanded about the mean and the first three terms are given below;

$$H(X) = H(\mu) + \frac{H'(\mu)}{1!}(X-\mu) + \frac{H''(\mu)}{2!}(X-\mu)^2 + \cdots$$

Where $H^n(\mu)$ is the nth derivative of H(X) evaluated at μ.

By applying the expectation on the above function;

$$E(H(X)) = E(H(\mu)) + E\left(H'(\mu)X\right) - E\left(H'(\mu)\mu\right) + \frac{H''(\mu)}{2}E(X-\mu)^2 + \cdots$$

Since the H (μ), $H'(\mu)$, and $H'(\mu)\mu$ are constants, and $E(X-\mu)^2$ is the variance of the random variable X the above equation results in the following form;

$$E(H(X)) = E(\mu) + H'(\mu)E(X) - H'(\mu)\mu + \frac{H''(\mu)}{2}V(X) + \cdots$$

$$E(H(X)) = H(\mu) + H'(\mu)\mu - H'(\mu)\mu + \frac{H''(\mu)}{2}\sigma^2 + \cdots$$

$$= H(\mu) + \frac{H''(\mu)}{2}\sigma^2 + \cdots$$

Then if the function of the random variable is H(X) the mean can be approximated using the following expression;

$$E(H(X)) \cong H(\mu) + \frac{1}{2}H''(\mu)\sigma^2$$

Similarly from the first two terms of the Taylor series expansion the variance of H(X) can be approximated with the following expression.

$$V(H(X)) \cong [H'(\mu)]^2 . \sigma^2$$

If the variance is large and the mean is small then it may be required to include more terms of the Taylor series in getting the mean and the variance.

Moment-generating Function

Moment generating functions are useful in finding the origin moments of a given random variable X. The moment-generating function of a random variable X is defined as follows;

$$M_x(t) = E(e^{tX})$$

$M_x(t) = \sum_{\forall i} e^{t_i x_i} px(x_i)$ is the moment generating function for a discrete random variable.

$M_x(t) = \int_{-\infty}^{\infty} e^{tx} fx(x)\,dx$ is the moment generating function for a continuous random variable.

The exponential function in the moment generating function can be expanded using Taylor series expansion as shown below.

$$e^{tX} = 1 + tX + \frac{t^2X^2}{2!} + \cdots + \frac{t^rX^r}{r!} + \cdots$$

Now, by finding the expectations of different terms one can get the moment generating function of the random variable.

$$M_x(t) = 1 + \mu_1 t + \mu_2 \frac{t^2}{2!} + \cdots + \mu_r \frac{t^r}{r!} + \cdots$$

From the above equation it can be seen that the coefficient of is the rth origin moment of the random variable. Computation of the moments of different commonly used probability distributions is described.

Multiple Random Variables

Definition

If Ω is the sample space associated to an experiment E and X_1 , X_2 , X_3 ,...,X_k are the random variables assigning the real values to the outcomes of the experiment, then [X_1 ,X_2 ,....X_k] is called a k-dimensional random vector.

Joint Distribution of two Random Variables

In this case, [X_1 , X_2] is a two dimensional random vector. If the values taken by this vector are integer then it is called a two dimensional discrete random vector and if it takes real numbers then it is called a two dimensional continuous random vector. The following are the joint probability mass and probability density functions for two dimensional discrete and continuous random vectors, respectively.

For discrete two dimensional random vector [X_1 , X_2], $f(x_1, x_2)$ = P($X_1 = x_1$ and $X_2 = x_2$) is the joint probability mass function.

For continuous two dimensional random vector there exists a function $f(x_1, x_2)$ such that

$$p(a_1 < X_1 < b_1, a_2 < X_2 < b_2) = \int_{a_1}^{b_2} \int_{a_2}^{b_2} f(x_1, x_2)\, dx_1 dx_2, and$$

$$\int_{-\infty}^{\infty} \int_{-\infty}^{\infty} f(x_1, x_2)\, dx_1 ds_2 = 1$$

This function is called as the joint probability density function. If the random variables X_1 and X_2 are stochastically independent, in case of discrete random variables the join mass function would be P(x_1) P(x_2). In case of continuous random variables the joint distribution function would be $f(x_1)f(x_2)$.

Problem 1

After widening the national highways from two to four lanes there is a significant increase in the number of accidents during night times. In majority of these cases it may be hypothesized that the accidents are occurring due to the on-road parking or due to vehicle breakdown. Let random variables X and Y denote the number of vehicles parked on-road as well as the breakdown vehicles (left on the road itself), and the number of accidents occurring over a 10 km road stretch, respectively. If the joint probability mass function of X and Y is as given in the following table, find the following.

1. Probability of exactly one on-road parking/breakdown and one accident per 10 km.

2. What is the probability of more than two accidents occurring over 10 km stretch?
3. Determine the marginal probability mass functions of X and Y.
4. Check whether X and Y are independent variables.

		Y				P(X)
		1	2	3	4	
X	0	0.05	0.01	0	0	0.06
	1	0.11	0.04	0.03	0.01	0.19
	2	0.14	0.06	0.04	0.05	0.29
	3	0.23	0.1	0.07	0.06	0.46
P(Y)		0.53	0.21	0.14	0.12	1

1. Probability of exactly one on-road parking/breakdown and one accident per 10 km.

This is nothing but P(X=1 and Y=1)
From the table cell entry for X=1 and Y=1, i.e. P(X=1 and Y=1), is 0.11.

2. What is the probability of more than two accidents occurring over 10 km stretch?

This is nothing but $P(Y > 2)$
$P(Y > 2) = P(Y = 3 \text{ and } X = 0, 1, 2, 3) + P(Y = 4 \text{ and } X = 0, 1, 2, 3)$
$= 0 + 0.03 + 0.04 + 0.07 + 0 + 0.01 + 0.05 + 0.06$
$= 0.26$

3. Determine the marginal probability mass functions of X and Y.

Probability mass function of X is

$$P(X = x) = \sum_{\forall y} p(x, y)$$

Random variable X takes 0, 1, 2, 3.

$$P(X=0) = \sum_{\forall y} p(0, y)$$
$$= 0.05 + 0.01 + 0 + 0$$
$$= 0.06$$

Similarly, for X = 1, p(X = 1) = 0.19; for X = 2, p(X = 2) = 0.29; for X = 3, p(X = 3) = 0.46. The same is shown as the last column of the combined probability mass function Table.

P(X)	0.06	0.19	0.29	0.46

Similarly, the marginal probability mass function of random variable Y is nothing but the last row of the combined probability mass function Table.

P(Y)	0.53	0.21	0.14	0.12

4.Check whether X and Y are independent variables.

If there exists any (x, y), the values taken by the random variables X and Y respectively, for which P(X=x and Y=y) is not equals to P(X = x) * P(Y = y), then the two random variables are not independent.
From the joint pmf table if X = 3 and Y=1, P(X = 3 and Y = 1) is 0.23.
p(X = 3) is 0.46 and p(Y = 1) is 0.53, hence P(X = 3) * P(Y = 1) is 0.24.

From the above calculations it can be said that X and Y are not independent.

Problem 2

The joint probability density function of two continuous random variables is as shown below.

$$f(x, y) = c * x * \exp[-x(1+y)] \quad for\ x\ and\ y \geq 0$$

1. Find the value of c?
2. What is the probability that the variable X takes values greater than 5?
3. Find the marginal probability density functions of X and Y.
4. Whether these two random variables are independent?

i. Find the value of c?

From the properties of the joint density functions it is known that

$$\int_0^\infty \int_0^\infty f(x, y)\, dxdy = 1$$

Then

$$\int_0^\infty \int_0^\infty c * x * \exp[-x(1+y)]\, dxdy = 1$$

$$c\int_0^\infty x \frac{\exp[-x(1+y)]}{-x} dx = 1$$

$$-c\int_0^\infty \exp[-x(1+y)] \Big|_0^\infty dx = 1$$

$$c\int_0^\infty e^{-x} dx = 1$$

$$-c\, e^{-x} \Big|_0^\infty = 1$$

$$-c(-1) = 1$$

Hence the constant c=1.

ii. What is the probability that the variable X takes values greater than 5?

$$p(5 < X < \infty, 0 < Y < \infty) = \int_S^{\infty}\int_0^{\infty} f(x,y)\,dy\,dx$$
$$= \int_S^{\infty}\int_0^{\infty} x \exp[-x(1+y)]dy\,dx$$
$$= \int_S^{\infty}\int_0^{\infty} e^{-x} dx$$
$$= -1 * e^{-x}\Big|_S^{\infty}$$
$$= -\left(-e^{-S}\right)$$
$$= 0.006738$$

iii. Find the marginal probability density functions of X and Y.

Marginal density function of the random variable X is

$$f(x) = \int_0^{\infty} f(x,y)\,dy$$
$$f(x) = \int_0^{\infty} x * \exp\left[-x(1+y)\right] dy$$
$$= e^{-x}$$

Marginal density function of the random variable Y is

$$f(y) = \int_0^{\infty} f(x,y)\,dx$$
$$f(y) = \int_0^{\infty} x * \exp\left[-x(1+y)\right] dx$$
$$= e^{-x}$$

$$f(y) = \frac{1}{(1+y)^2}\begin{cases} \textit{from} \int u dv = uv - \int v du \\ u = x \\ \textit{and } dv = \exp\left[-x(1+y)\right]dx \end{cases}$$

iv. Whether these two random variables are independent?

If the random variables X and Y are independent the joint probability density function will be equals to the product of the marginal density functions.

$$f(x,y) = f(x) * f(y)$$

From the answer for the third part of the question it can be seen that the random variables X and Y are not independent.

Conditional Probability Functions

Among the two random variables X_1 and X_2, if it is to find the mass or distribution function of one random variable, say X_1, at a particular value of the other random variable the concept of conditional probability can be used.

If the random variables X_1 and X_2 are discrete, the probability mass function of X_1 at a given value of $X_2 = x_2$ will be as given below;

$$p\left[X_1 = x / X_2 = x_2\right] = \frac{p\left[(X_1 = x)\cap(X_2 = x_2)\right]}{p\left[X_2 = x_2\right]}$$

$$= \frac{p(x, x_2)}{p\left[X_2 = x_2\right]}$$

$$= \frac{p(x, x_2)}{\sum_{\forall x_i} p(x_i, x_2)}$$

$$= \frac{p(x_i, x_2)}{p(x_2)}$$

If X_1, and X_2 are stochastically independent random variables the above expression would become;

$$p\left[X_1 = x / X_2 = x_2\right] = \frac{p\left[(X_1 = x)\cap(X_2 = x_2)\right]}{p\left[X_2 = x_2\right]}$$

$$= \frac{p\left[(X_1 = x) * p(X_2 = x_2)\right]}{p\left[X_2 = x_2\right]}$$

$$= p(x_1)$$

Similarly, if X_1 and X_2 are two continuous random variables, the conditional probability density function of X_1 for a given X_2 would be,

$$f\left(X_1 / X_2\right) = \frac{f(x_1, x_2)}{f(x_2)}$$

If X_1 and X_2 are two stochastically independent continuous random variables then the above expression for the conditional probability density function would become;

$$f\left(x_1 / x_2\right) = \frac{f(x_1, x_2)}{f(x_2)}$$

$$= f(x_1)$$

Marginal Distributions in Case of two Dimensional Random Vectors

If [X_1, X_2], is the two dimensional random vector, sometimes it is necessary to know the distribution of eitherX_1 or X_2. These are called as marginal distributions. In case of a discrete two dimensional random vector the marginal distribution of X_1 is given below;

$$P(X_1 = x_1) = \sum_{\forall x_2} p(x_1, x_2)$$

In case of a continuous two dimensional random vector the marginal distribution of X_1 is given below;

$$f_1(x_1) = \int_{-\infty}^{\infty} f(x_1, x_2)\, dx_2$$

The procedure adopted in obtaining the mean and variance of the marginal distribution is similar to the procedure adopted in the previous sections.

Conditional Expectation

If X and Y are the discrete random variables and the joint probability mass function of these two random variables is p(X=x and Y=y), the conditional expectation of h(X) given Y is

$$E(h(X)/Y) = \sum_{for\ all\ x} h(x)\, p\left(x/_y\right)$$

If X and Y are the continuous random variables and the joint probability density function of these two random variables is f(x,y), the conditional expectation of h(X) given Y is

$$E(h(X)/Y) = \int_{-\infty}^{\infty} h(x)\, f\left(x/_y\right) dx$$

Functions of More than one Random Variable

When using the probability theory in engineering applications often it is necessary to deal with simultaneously more than one random variable. For example, in traffic engineering applications, more specifically in simulation models, it is necessary to generate vehicles and this process involves more than one random variable. This is a typical example for a bi-variate distribution. This is explained in detail when dealing with the composite distributions for modeling the headways. It is also important to understand these functions as they are useful when dealing with the random samples which are composed of several independent random variables. Sometimes they are identical and sometimes not. It is necessary to understand these functions to formulate joint probability distributions and associated marginal and conditional distributions.

Problem 1

Using the following joint probability mass function it is known that the random variables X and Y are not independent. X and Y denote the number of vehicles parked on-road as well as the break-down vehicles, and the number of accidents occurring over a 10 km road stretch, respectively.

		Y				P(X)
		1	2	3	4	
X	0	0.05	0.01	0	0	0.06
	1	0.11	0.04	0.03	0.01	0.19
	2	0.14	0.06	0.04	0.05	0.29
	3	0.23	0.1	0.07	0.06	0.46
P(Y)		0.53	0.21	0.14	0.12	1

From the joint probability mass function, find the following;

1. What is the probability of the number of accidents occurring over 10 km road stretch equals to 1 given that the number of on-road parked/breakdown vehicles over the same road stretch is 2?

It can be expressed as

$$p\left(\frac{Y=1}{X=2}\right)=\frac{p(Y=y \; and \; X=2)}{P(X=2)}$$
$$=\frac{p(Y=1 \; and \; X=2)}{P(X=2)}$$
$$=\frac{.14}{0.29}$$
$$=0.483$$

2. What is the probability of the number of accidents less than or equals to 3 given the number of on-road parked/breakdown vehicles over the same road stretch is 2?

It can be expressed as

$$p\left(\frac{Y\le 3}{X=2}\right)=\frac{p(Y\le 3 \; and \; X=2)}{P(X=2)}$$
$$=\frac{p(Y=1 \; and \; X=2)+p(Y=2 \; and \; X=2)+p(Y=3 \; and \; X=2)}{P(X=2)}$$
$$=\frac{0.14+0.06+0.04}{0.29}$$
$$=0.83$$

Problem 2

The joint probability density function of two continuous random variables is as shown below;

$$f(x,y) = x * \exp\left[-x(1+y)\right] \; for \; x \; and \; y \geq 0$$

1. Find whether the random variables are independent or not.
2. Find the conditional probability density function of X given Y.
3. Find the conditional probability density function of Y given X.
4. Find the conditional expectation of X given Y.

- Find whether the random variables are independent or not.

From the example it is known that the two random variables are not independent.

- Find the conditional probability density function of X given Y.

Conditional probability density function of the random variable X is

$$f\left(x/y\right) = \frac{f(x,y)}{f(y)}$$

The numerator is the joint probability density function. The denominator is the marginal density function of the random variable Y.
Marginal density function of Y is

$$f(y) = \frac{1}{(1+y)^2}$$

Hence the conditional probability density function of X is;

$$f\left(x/y\right) = \frac{x * \exp\left[-x(1+y)\right]}{\frac{1}{(1+y)^2}}$$
$$= (1+y)^2 * x * \exp\left[-x(1+y)\right] \; for \; x \; and \; y \geq 0$$

- Find the conditional probability density function of Y given X.

Marginal density function of X is

$$f(x) = \int_0^\infty f(x,y)\, dy$$
$$f(x) = \int_0^\infty x * \exp\left[-x(1+y)\right] dy$$
$$= e^{-x}$$

Hence the conditional probability density function of X is;

$$f\left(\frac{y}{x}\right)=\frac{x*\exp\left[-x(1+y)\right]}{e^{-x}}$$
$$=x*\exp\left[-x(1+y)\right]*e^{x}$$
$$=x*e^{-xy} \qquad for\ x\ and\ y\geq 0$$

Problem 3

X and Y are two continuous random variables and have the following joint probability density function.

$$f(x,y)=\exp\left[-(x+y)\right]\ for\ x\ and\ y\geq 0$$

Find the following;

1.Conditional expectation of X given Y.

As a first step to get the conditional expectation the conditional density functions are to be obtained.

$$f(x/y)=\frac{f(x,y)}{f(y)}$$

Once the marginal density functions are obtained it can be seen that the two random variables are independent. Then,

$$f\left(\frac{x}{y}\right)=\frac{f(x)f(y)}{f(y)}$$
$$=f(x)$$

The conditional expectation of X given Y is,

$$E(X/Y)=\int_0^\infty xf\left(\frac{x}{y}\right)dx$$
$$=\int_0^\infty xf(x)\,dx$$
$$=\int_0^\infty x\exp[-x]\,dx$$
$$=-x\exp[-x]\Big|_0^\infty+\int_0^\infty \exp[-x]\,dx$$
$$=1$$

2. Conditional expectation of Y given X

$$f(y/x)=\frac{f(x,y)}{f(x)}$$

Since X and Y are independent random variables;

$$f(y/x)=f(y)$$

The conditional expectation of Y given X is,

$$E(Y/K)=\int_{0}^{\infty} yf\left(\frac{y}{x}\right)dy$$

$$=\int_{0}^{\infty} yf(y)dy$$

$$=\int_{0}^{\infty} y\exp(-y)dy$$

$$=1$$

4

A Comprehensive Study of Probability Distribution

Probability distribution is mathematical function that provides the set of all the possible outcomes of an experiment. It can be divided into two parts, namely discrete probability distribution and continuous probability distribution. While the former deals with situations which have a finite number of outcomes, the latter is related to scenarios that have an infinite number of possibilities. The major components of probability distribution are discussed in this chapter.

Probability Distribution

In probability theory and statistics, a probability distribution is a mathematical function that, stated in simple terms, can be thought of as providing the probability of occurrence of different possible outcomes in an experiment. For instance, if the random variable X is used to denote the outcome of a coin toss ('the experiment'), then the probability distribution of X would take the value 0.5 for $X = \text{heads}$, and 0.5 for $X = \text{tails}$.

In more technical terms, the probability distribution is a description of a random phenomenon in terms of the probabilities of events. Examples of random phenomena can include the results of an experiment or survey. A probability distribution is defined in terms of an underlying sample space, which is the set of all possible outcomes of the random phenomenon being observed. The sample space may be the set of real numbers or a higher-dimensional vector space, or it may be a list of non-numerical values; for example, the sample space of a coin flip would be {heads, tails}.

Probability distributions are generally divided into two classes. A discrete probability distribution (applicable to the scenario where the set of possible outcomes is discrete, such as a coin toss or a roll of dice) can be encoded by a discrete list of the probabilities of the outcomes, known as a probability mass function. On the other hand, a continuous probability distribution (applicable to the scenarios where the set of possible outcomes can take on values in a continuous range (e.g., real numbers), such as the temperature on a given day) is typically described by probability density functions (with the probability of any individual outcome actually being 0). The normal distribution represents a commonly encountered continuous probability distribution. More complex experiments, such as those involving stochastic processes defined in continuous time, may demand the use of more general probability measures.

A probability distribution whose sample space is the set of real numbers is called univariate, while a distribution whose sample space is a vector space is called multivariate. A univariate distribution gives the probabilities of a single random variable taking on various alternative values; a multivariate distribution (a joint probability distribution) gives the probabilities of a random vector—a list of two or more random variables—taking on various combinations of values. Important and commonly encountered univariate probability distributions include the binomial distribution, the hypergeometric distribution, and the normal distribution. The multivariate normal distribution is a commonly encountered multivariate distribution.

Introduction

To define probability distributions for the simplest cases, one needs to distinguish between discrete and continuous random variables. In the discrete case, it is sufficient to specify a probability mass function assigning a probability to each possible outcome: for example, when throwing a fair dice, each of the six values *1* to *6* has the probability 1/6. The probability of an event is then defined to be the sum of the probabilities of the outcomes that satisfy the event; for example, the probability of the event "the die rolls an even value" is $\text{Prob}(2)+\text{Prob}(4)+\text{Prob}(6)=1/6+1/6+1/6=1/2$.

In contrast, when a random variable takes values from a continuum then typically, any individual outcome has probability zero and only events that include infinitely many outcomes, such as intervals, can have positive probability. For example, the probability that a given object weighs *exactly* 500 g is zero, because the probability of measuring exactly 500 g tends to zero as the accuracy of our measuring instruments increases. Nevertheless, in quality control one might demand that the probability of a "500 g" package containing between 490 g and 510 g should be no less than 98%, and this demand is less sensitive to the accuracy of our instruments.

Continuous probability distributions can be described in several ways. The probability density function describes the infinitesimal probability of any given value, and the probability that the outcome lies in a given interval can be computed by integrating the probability density function over that interval. On the other hand, the cumulative distribution function describes the probability that the random variable is no larger than a given value; the probability that the outcome lies in a given interval can be computed by taking the difference between the values of the cumulative distribution function at the endpoints of the interval. The cumulative distribution function is the antiderivative of the probability density function provided that the latter function exists.

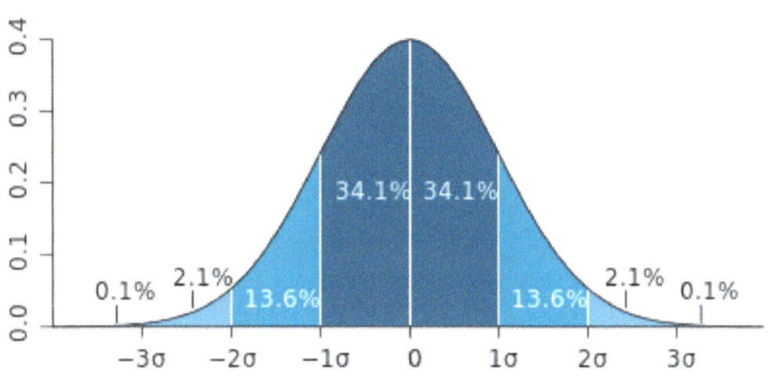

The probability density function (pdf) of the normal distribution, also called Gaussian or "bell curve", the most important continuous random distribution. As notated on the figure, the probabilities of intervals of values correspond to the area under the curve.

Terminology

As probability theory is used in quite diverse applications, terminology is not uniform and sometimes confusing. The following terms are used for non-cumulative probability distribution functions:

- Distribution, Frequency distribution: is a table that displays the frequency of various outcomes in a sample.
- Probability distribution: is a table that displays the probabilities of various outcomes in a sample. Could be called a "normalized frequency distribution table", where all occurrences of outcomes sum to 1.
- Distribution function: is a functional form of frequency distribution table.
- Probability distribution function: is a functional form of probability distribution table. Could be called a "normalized frequency distribution function", where area under the graph equals to 1.

Finally,

- Probability mass, Probability mass function, p.m.f., Discrete probability distribution function: for discrete random variables.
- Categorical distribution: for discrete random variables with a finite set of values.
- Probability density, Probability density function, p.d.f., Continuous probability distribution function: most often reserved for continuous random variables.

The following terms are somewhat ambiguous as they can refer to non-cumulative or cumulative distributions, depending on authors' preferences:

- Probability distribution function: continuous or discrete, non-cumulative or cumulative.
- Probability function: even more ambiguous, can mean any of the above or other things.

Basic Terms

- Mode: for a discrete random variable, the value with highest probability (the location at which the probability mass function has its peak); for a continuous random variable, the location at which the probability density function has its peak.
- Support: the smallest closed set whose complement has probability zero.
- Head: the range of values where the pmf or pdf is relatively high.
- Tail: the complement of the head within the support; the large set of values where the pmf or pdf is relatively low.
- Expected value or mean: the weighted average of the possible values, using their probabilities as their weights; or the continuous analog thereof.
- Median: the value such that the set of values less than the median has a probability of one-half.
- Variance: the second moment of the pmf or pdf about the mean; an important measure of the dispersion of the distribution.
- Standard deviation: the square root of the variance, and hence another measure of dispersion.
- Symmetry: a property of some distributions in which the portion of the distribution to the left of a specific value is a mirror image of the portion to its right.

- Skewness: a measure of the extent to which a pmf or pdf "leans" to one side of its mean. The third standardized moment of the distribution.
- Kurtosis: a measure of the "fatness" of the tails of a pmf or pdf. The fourth standardized moment of the distribution.

Cumulative Distribution Function

Because a probability distribution Pr on the real line is determined by the probability of a scalar random variable X being in a half-open interval $(-\infty, x]$, the probability distribution is completely characterized by its cumulative distribution function:

$$F(x) = \Pr[X \leq x] \qquad \text{for all } x \in \mathbb{R}.$$

Discrete Probability Distribution

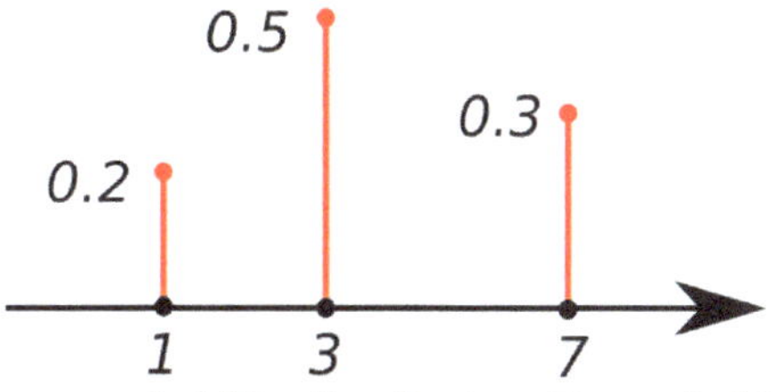

The probability mass function of a discrete probability distribution. The probabilities of the singletons {1}, {3}, and {7} are respectively 0.2, 0.5, 0.3. A set not containing any of these points has probability zero.

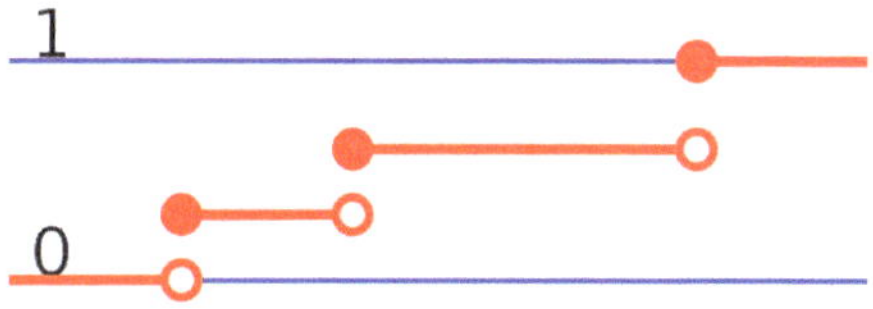

The cdf of a discrete probability distribution, ...

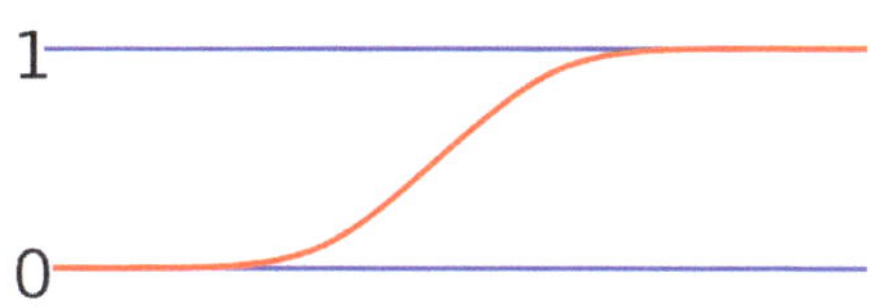

... of a continuous probability distribution, ...

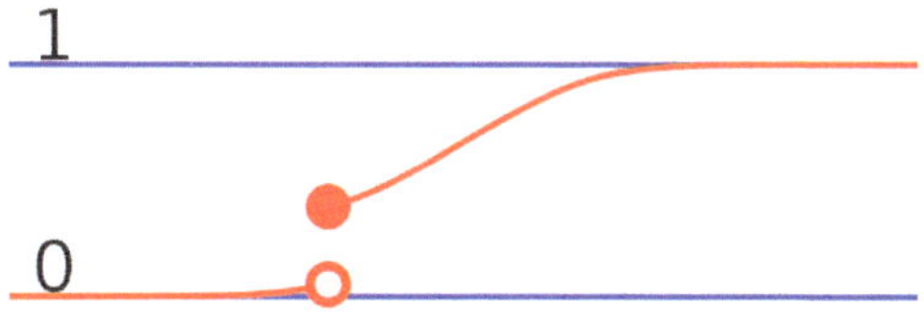

... of a distribution which has both a continuous part and a discrete part.

A discrete probability distribution is a probability distribution characterized by a probability mass function. Thus, the distribution of a random variable X is discrete, and X is called a discrete random variable, if

$$\sum_u \Pr(X = u) = 1$$

as u runs through the set of all possible values of X. A discrete random variable can assume only a finite or countably infinite number of values. For the number of potential values to be countably infinite, even though their probabilities sum to 1, the probabilities have to decline to zero fast enough. For example, if $\Pr(X = n) = \frac{1}{2^n}$ for n = 1, 2, ..., we have the sum of probabilities 1/2 + 1/4 + 1/8 + ... = 1.

Well-known discrete probability distributions used in statistical modeling include the Poisson distribution, the Bernoulli distribution, the binomial distribution, the geometric distribution, and the negative binomial distribution. Additionally, the discrete uniform distribution is commonly used in computer programs that make equal-probability random selections between a number of choices.

Measure Theoretic Formulation

A measurable function $X : A \to B$ between a probability space $(A, \mathcal{A}, P)$ and a measurable space $(B, \mathcal{B})$ is called a discrete random variable provided its image is a countable set and the pre-image of singleton sets are measurable, i.e., $X^{-1}(b) \in \mathcal{A}$ for all $b \in B$. The latter requirement induces a probability mass function $f_X : X(A) \to \mathbb{R}$ via $f_X(b) := P(X^{-1}(b))$. Since the pre-images of disjoint sets are disjoint

$$\sum_{b \in X(A)} f_X(b) = \sum_{b \in X(A)} P(X^{-1}(b)) = P\left(\bigcup_{b \in X(A)} X^{-1}(b)\right) = P(A) = 1.$$

This recovers the definition given above.

Cumulative Density

Equivalently to the above, a discrete random variable can be defined as a random variable whose cumulative distribution function (cdf) increases only by jump discontinuities—that is, its cdf increases only where it "jumps" to a higher value, and is constant between those jumps. The points where jumps occur are precisely the values which the random variable may take.

Delta-function Representation

Consequently, a discrete probability distribution is often represented as a generalized probability density function involving Dirac delta functions, which substantially unifies the treatment of continuous and discrete distributions. This is especially useful when dealing with probability distributions involving both a continuous and a discrete part.

Indicator-function Representation

For a discrete random variable X, let u_0, u_1, ... be the values it can take with non-zero probability. Denote

$$\Omega_i = X^{-1}(u_i) = \{\omega : X(\omega) = u_i\}, i = 0,1,2,\dots$$

These are disjoint sets, and by formula (1)

$$\Pr\left(\bigcup_i \Omega_i\right) = \sum_i \Pr(\Omega_i) = \sum_i \Pr(X = u_i) = 1.$$

It follows that the probability that X takes any value except for u_0, u_1, ... is zero, and thus one can write X as

$$X(\omega) = \sum_i u_i 1_{\Omega_i}(\omega)$$

except on a set of probability zero, where 1_A is the indicator function of A. This may serve as an alternative definition of discrete random variables.

Continuous Probability Distribution

A continuous probability distribution is a *probability distribution* that has a cumulative distribution function that is continuous. Most often they are generated by having a probability density function. Mathematicians call distributions with probability density functions absolutely continuous, since their cumulative distribution function is absolutely continuous with respect to the Lebesgue measure λ. If the distribution of X is continuous, then X is called a continuous random variable. There are many examples of continuous probability distributions: normal, uniform, chi-squared, and others.

Intuitively, a continuous random variable is the one which can take a continuous range of values—as opposed to a discrete distribution, where the set of possible values for the random variable is at most countable. While for a discrete distribution an event with probability zero is impossible (e.g., rolling 31/2 on a standard dice is impossible, and has probability zero), this is not so in the case of a continuous random variable. For example, if one measures the width of an oak leaf, the result of 3½ cm is possible; however, it has probability zero because uncountably many other potential values exist even between 3 cm and 4 cm. Each of these individual outcomes has probability zero, yet the probability that the outcome will fall into the interval (3 cm, 4 cm) is nonzero. This apparent paradox is resolved by the fact that the probability that X attains some value within an infinite set, such as an interval, cannot be found by naively adding the probabilities for individual values. Formally, each value has an infinitesimally small probability, which statistically is equivalent to zero.

Formally, if X is a continuous random variable, then it has a probability density function $f(x)$, and therefore its probability of falling into a given interval, say $[a, b]$ is given by the integral

$$\Pr[a \le X \le b] = \int_a^b f(x)dx$$

In particular, the probability for X to take any single value a (that is $a \le X \le a$) is zero, because an integral with coinciding upper and lower limits is always equal to zero.

The definition states that a continuous probability distribution must possess a density, or equivalently, its cumulative distribution function be absolutely continuous. This requirement is stronger than simple continuity of the cumulative distribution function, and there is a special class of distributions, singular distributions, which are neither continuous nor discrete nor a mixture of those. An example is given by the Cantor distribution. Such singular distributions however are never encountered in practice.

Note on terminology: some authors use the term "continuous distribution" to denote the distribution with continuous cumulative distribution function. Thus, their definition includes both the (absolutely) continuous and singular distributions.

By one convention, a probability distribution μ is called *continuous* if its cumulative distribution function $F(x) = \mu(-\infty, x]$ is continuous and, therefore, the probability measure of singletons $\mu\{x\}=0$ for all x.

Another convention reserves the term *continuous probability distribution* for absolutely continuous distributions. These distributions can be characterized by a probability density function: a non-negative Lebesgue integrable function f defined on the real numbers such that

$$F(x) = \mu(-\infty, x] = \int_{-\infty}^{x} f(t)dt.$$

Discrete distributions and some continuous distributions (like the Cantor distribution) do not admit such a density.

Some Properties

- The probability distribution of the sum of two independent random variables is the convolution of each of their distributions.
- Probability distributions are not a vector space—they are not closed under linear combinations, as these do not preserve non-negativity or total integral 1—but they are closed under convex combination, thus forming a convex subset of the space of functions (or measures).

Kolmogorov Definition

In the measure-theoretic formalization of probability theory, a random variable is defined as a measurable function X from a probability space $(\Omega, \mathcal{F}, \mathrm{P})$ to measurable space $(\mathcal{X}, \mathcal{A})$. A probability distribution of $\boldsymbol{X}$ is the pushforward measure $X_*\mathrm{P}$ of X, which is a probability measure on $(\mathcal{X}, \mathcal{A})$ satisfying $X_*\mathrm{P} = \mathrm{P}X^{-1}$.

Random Number Generation

A frequent problem in statistical simulations (the Monte Carlo method) is the generation of pseudo-random numbers that are distributed in a given way. Most algorithms are based on a pseudorandom number generator that produces numbers X that are uniformly distributed in the interval [0,1]. These random variates X are then transformed via some algorithm to create a new random variate having the required probability distribution.

Applications

The concept of the probability distribution and the random variables which they describe underlies the mathematical discipline of probability theory, and the science of statistics. There is spread or variability in almost any value that can be measured in a population (e.g. height of people, durability of a metal, sales growth, traffic flow, etc.); almost all measurements are made with some intrinsic error; in physics many processes are described probabilistically, from the kinetic properties of gases to the quantum mechanical description of fundamental particles. For these and many other reasons, simple numbers are often inadequate for describing a quantity, while probability distributions are often more appropriate.

As a more specific example of an application, the cache language models and other statistical language models used in natural language processing to assign probabilities to the occurrence of particular words and word sequences do so by means of probability distributions.

Common Probability Distributions

The following is a list of some of the most common probability distributions, grouped by the type of process that they are related to.

Note also that all of the univariate distributions below are singly peaked; that is, it is assumed that the values cluster around a single point. In practice, actually observed quantities may cluster around multiple values. Such quantities can be modeled using a mixture distribution.

Related to Real-valued Quantities that Grow Linearly (e.g. Errors, Offsets)

- Normal distribution (Gaussian distribution), for a single such quantity; the most common continuous distribution

Related to Positive real-valued Quantities that Grow Exponentially (e.g. Prices, Incomes, Populations)

- Log-normal distribution, for a single such quantity whose log is normally distributed
- Pareto distribution, for a single such quantity whose log is exponentially distributed; the prototypical power law distribution

Related to Real-valued Quantities that are Assumed to be Uniformly Distributed over a (Possibly Unknown) Region

- Discrete uniform distribution, for a finite set of values (e.g. the outcome of a fair die)
- Continuous uniform distribution, for continuously distributed values

Related to Bernoulli Trials (Yes/No Events, with a Given Probability)

- Basic distributions:
    - Bernoulli distribution, for the outcome of a single Bernoulli trial (e.g. success/failure, yes/no)

  - Binomial distribution, for the number of "positive occurrences" (e.g. successes, yes votes, etc.) given a fixed total number of independent occurrences
  - Negative binomial distribution, for binomial-type observations but where the quantity of interest is the number of failures before a given number of successes occurs
  - Geometric distribution, for binomial-type observations but where the quantity of interest is the number of failures before the first success; a special case of the negative binomial distribution

- Related to sampling schemes over a finite population:
  - Hypergeometric distribution, for the number of "positive occurrences" (e.g. successes, yes votes, etc.) given a fixed number of total occurrences, using sampling without replacement
  - Beta-binomial distribution, for the number of "positive occurrences" (e.g. successes, yes votes, etc.) given a fixed number of total occurrences, sampling using a Polya urn scheme (in some sense, the "opposite" of sampling without replacement)

Related to Categorical Outcomes (Events with K Possible Outcomes, with a given Probability for Each Outcome)

- Categorical distribution, for a single categorical outcome (e.g. yes/no/maybe in a survey); a generalization of the Bernoulli distribution
- Multinomial distribution, for the number of each type of categorical outcome, given a fixed number of total outcomes; a generalization of the binomial distribution
- Multivariate hypergeometric distribution, similar to the multinomial distribution, but using sampling without replacement; a generalization of the hypergeometric distribution

Related to Events in a Poisson Process (Events that Occur Independently with a Given Rate)

- Poisson distribution, for the number of occurrences of a Poisson-type event in a given period of time
- Exponential distribution, for the time before the next Poisson-type event occurs
- Gamma distribution, for the time before the next k Poisson-type events occur

Related to the Absolute Values of Vectors with Normally Distributed Components

- Rayleigh distribution, for the distribution of vector magnitudes with Gaussian distributed orthogonal components. Rayleigh distributions are found in RF signals with Gaussian real and imaginary components.
- Rice distribution, a generalization of the Rayleigh distributions for where there is a stationary background signal component. Found in Rician fading of radio signals due to multipath

propagation and in MR images with noise corruption on non-zero NMR signals.

Related to Normally Distributed Quantities Operated with Sum of Squares (for Hypothesis Testing)

- Chi-squared distribution, the distribution of a sum of squared standard normal variables; useful e.g. for inference regarding the sample variance of normally distributed samples
- Student's t distribution, the distribution of the ratio of a standard normal variable and the square root of a scaled chi squared variable; useful for inference regarding the mean of normally distributed samples with unknown variance
- F-distribution, the distribution of the ratio of two scaled chi squared variables; useful e.g. for inferences that involve comparing variances or involving R-squared (the squared correlation coefficient)

Useful as Conjugate Prior Distributions in Bayesian Inference

- Beta distribution, for a single probability (real number between 0 and 1); conjugate to the Bernoulli distribution and binomial distribution
- Gamma distribution, for a non-negative scaling parameter; conjugate to the rate parameter of a Poisson distribution or exponential distribution, the precision (inverse variance) of a normal distribution, etc.
- Dirichlet distribution, for a vector of probabilities that must sum to 1; conjugate to the categorical distribution and multinomial distribution; generalization of the beta distribution
- Wishart distribution, for a symmetric non-negative definite matrix; conjugate to the inverse of the covariance matrix of a multivariate normal distribution; generalization of the gamma distribution

Discrete Probability Distributions

If the experimental outcome or the data from the natural processes takes on finite number of values, the data may fit into any of the following discrete distributions.

The Bernoulli Distribution

The Bernoulli distribution applies to the experiments that have only two possible outcomes, often a "success" S and a "failure" F. Tossing a fair coin, fair or unfair, is a Bernoulli trial. In any trial the probability of success, *p*, and the probability of failure, *q*, is given by the following equation;

$$p[F] = q = 1 - P[S] = 1 - P$$

The probability mass function for a Bernoulli random variable is;

$$p(x) = p^x(1-p)^{1-x} \text{ for } x = 0, 1 \text{ and } 0 \le p \le 1$$

p(x) = 0 otherwise.

In the special case when the success is coded as 1 and the failure as 0, then:

The mean of a Bernoulli variable is,

$$E(X_j) = (0.q) + (1.p) = p$$

Variance of a Bernoulli variable is

$$V(X_j) = [(0^2.q) + (1^2.p)] - p^2 = p(1-p) = pq$$

Probability mass function for a Bernoulli random variable with n=25 and p = 0.2 is shown in the figure below.

Applications

Though direct applications of Bernoulli distribution are not there, the distribution forms taken up by the sequence of Bernoulli trials have several applications in Civil Engineering.

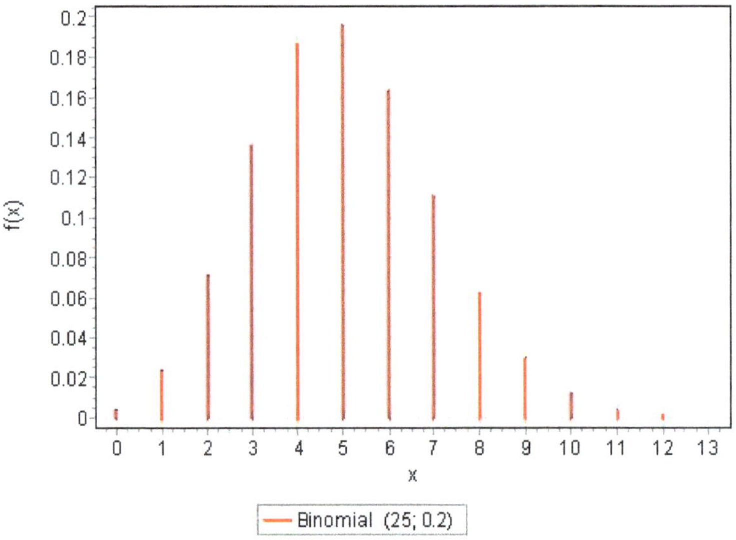

Probability mass function of a Binomial random variable with n=25 and p 0.2

The Binomial Distribution

The binomial distribution expresses the probability of x successes in a sequence of n independent Bernoulli trials.

$$P(x) = P[X = x] = \frac{n!}{x!(n-x)!} p^x q^{n-x}$$

The mean and the variance of the binomial distribution are:

Mean of a binomial variable

$$E(x)=\sum_{x=0}^{n} x.\frac{n!}{x!(n-x)!}p^x q^{n-x}$$
$$=np\sum_{x=1}^{n}\frac{(n-1)!}{(x-1)!(n-x)}p^{x-1}q^{n-x}$$

And letting $y = x - 1$

$$E(X)=np\sum_{x=0}^{n-1}\frac{(n-1)!}{y!(n-1-y)!}p^y q^{n-1-y}$$

Since the summation in the above equation is equals to unity,

$$E(X)=np$$

Variance of a binomial variable

$$V(X)=E(X^2)-(E(X))^2$$
$$=E(X^2-X+X)-(E(X))^2$$
$$=E(X(X-1)+X)-(E(X))^2$$
$$=E(X(X-1))+E(X)-(E(X))^2$$
$$=\sum_{x=0}^{n} x(x-1)\frac{n!}{x!(n-x)!}p^x q^{n-x}+np-(np)^2$$
$$=\sum_{x=2}^{n} x(x-1)\frac{n(n-1)(n-2)!}{x(x-1)(x-2)!(n-x)!}p^2 p^{x-2}q^{n-x}+np-(np)^2$$
$$=n(n-1)p^2\sum_{x=2}^{n}\frac{(n-2)!}{(x-2)!(n-x)!}p^{x-2}q^{n-x}+np-(np)^2$$

Since the summation in the above equation is equals to unity,

$$V(X)=npq$$

The Bernoulli distribution is a particular case of binomial distribution when n=1.

Applications of Binomial Distribution

In modeling the driver behavior, intersection turning movements, and in speed studies this distribution is used. This has several applications in other fields of civil engineering, such as the probability of occurrence of peak floods greater than the design peak flood in a particular time period, probability of peak ground acceleration exceeding certain design value in a given time interval etc.

For example, if the probability of a vehicle turning left at an intersection is 0.15 then the probability of 3 vehicles out of 10 vehicles turning left equals to,

$$^{10}C_3(0.15)^3(0.85)^7 = 0.130$$

In the above example, a specific vehicle turning left or not is a Bernoulli trial and it is assumed that the arrivals of individual vehicles at the junction are independent events.

Problem

Concrete block pavements are widely being used in parking lots, footpaths, and other public places. These blocks must have minimum strength and are manufactured accordingly. Due to the variations in the mix proportions all the blocks may not satisfy the strength requirements. When a strength test is conducted on this block it may fail or pass the test and this may be considered as Bernoulli trial. From many tests it has been observed that the probability of failure is 0.05. When a sample of 20 cubes are taken what is the probability of having exactly 2 cubes that may fail in the test?

Probability of passing the strength test, p = 0.95

Failing the test, q = 0.05

n = 20

Since a test on the concrete block is a Bernoulli trial the sequence of Bernoulli trials are assumed to follow binomial distribution. (It is assumed that the probability of failing/passing the test remains same for all the blocks)

Probability of having two cubes failing in the test is equivalent to probability of 18 cubes passing the test, then,

$$p(X = 18) = 20c_{18}0.95^{18}0.05^2$$
$$= 0.189$$

Geometric Distribution

In probability theory and statistics, the geometric distribution is either of two discrete probability distributions:

- The probability distribution of the number *X* of Bernoulli trials needed to get one success, supported on the set { 1, 2, 3, ...}
- The probability distribution of the number *Y* = *X* – 1 of failures before the first success, supported on the set { 0, 1, 2, 3, ... }

Which of these one calls "the" geometric distribution is a matter of convention and convenience.

These two different geometric distributions should not be confused with each other. Often, the name *shifted* geometric distribution is adopted for the former one (distribution of the number X); however, to avoid ambiguity, it is considered wise to indicate which is intended, by mentioning the support explicitly.

It's the probability that the first occurrence of success requires k number of independent trials, each with success probability p. If the probability of success on each trial is p, then the probability that the kth trial (out of k trials) is the first success is

$$\Pr(X = k) = (1-p)^{k-1}\,p$$

for k = 1, 2, 3, ...

The above form of geometric distribution is used for modeling the number of trials up to and including the first success. By contrast, the following form of the geometric distribution is used for modeling the number of failures until the first success:

$$\Pr(Y = k) = (1-p)^{k}\,p$$

for k = 0, 1, 2, 3, ...

In either case, the sequence of probabilities is a geometric sequence.

For example, suppose an ordinary die is thrown repeatedly until the first time a "1" appears. The probability distribution of the number of times it is thrown is supported on the infinite set { 1, 2, 3, ... } and is a geometric distribution with $p = 1/6$.

Introduction to the Geometric Distribution

Consider a sequence of trials, where each trial has only two possible outcomes (designated failure and success). The probability of success is assumed to be the same for each trial. In such a sequence of trials, the geometric distribution is useful to model the number of failures before the first success. The distribution gives the probability that there are zero failures before the first success, one failure before the first success, two failures before the first success, and so on.

Examples

A newly-wed couple plans to have children, and will continue until the first girl. What is the probability that there are zero boys before the first girl, one boy before the first girl, two boys before the first girl, and so on?

A doctor is seeking an anti-depressant for a newly diagnosed patient. Suppose that, of the available anti-depressant drugs, the probability that any particular drug will be effective for a particular patient is p=0.6. What is the probability that the first drug found to be effective for this patient is the first drug tried, the second drug tried, and so on? What is the expected number of drugs that will be tried to find one that is effective?

A patient is waiting for a suitable matching kidney donor for a transplant. If the probability that a randomly selected donor is a suitable match is p=0.1, what is the expected number of donors who will be tested before a matching donor is found?

Assumptions: When is the Geometric Distribution an Appropriate Model?

The geometric distribution is an appropriate model if the following assumptions are true.

- The phenomenon being modelled is a sequence of independent trials.
- There are only two possible outcomes for each trial, often designated success or failure.
- The probability of success, p, is the same for every trial.

If these conditions are true, then the geometric random variable is the count of the number of failures before the first success. The possible number of failures before the first success is 0, 1, 2, 3, and so on. The geometric random variable Y is the number of failures before the first success. In the graphs above, this formulation is shown on the right.

An alternative formulation is that the geometric random variable X is the total number of trials up to and including the first success, and the number of failures is X-1. In the graphs above, this formulation is shown on the left.

Probability of Outcomes

Consider the anti-depressant example above. The probability that any given drug is effective (success) is $p = 0.6$. The probability that a drug will not be effective (fail) is $q = 1 - p = 1 - 0.6 = 0.4$. Here are probabilities of some possible outcomes.

(i) The first drug works. There are zero failures before the first success. Y = 0 failures. The probability P(zero failures before first success) is simply the probability that the first drug works.

$$\Pr(Y = 0) = q^0 p = 0.4^0 \times 0.6 = 1 \times 0.6 = 0.6.$$

(ii) The first drug fails, but the second drug works. There is one failure before the first success. Y= 1 failure. The probability for this sequence of events is p(first drug fails) $\times$p(second drug is success) which is given by

$$\Pr(Y = 1) = q^1 p = 0.4^1 \times 0.6 = 0.4 \times 0.6 = 0.24.$$

(iii) The first drug fails, the second drug fails, but the third drug works. There are two failures before the first success. Y= 2 failures. The probability for this sequence of events is p(first drug fails) $\times$p(second drug fails) $\times$p(third drug is success)

$$\Pr(Y = 2) = q^2 p, = 0.4^2 \times 0.6 = 0.096.$$

The general formula to calculate the probability of k failures before the first success, where the probability of success is p and the probability of failure is $q = 1 - p$, is

$$\Pr(Y = k) = q^k p.$$

for $k = 0, 1, 2, 3, \ldots$.

For the newly-weds awaiting their first girl, the probability of no boys before the first girl is

$$\Pr(Y = 0) = q^0 p = 0.5^0 \times 0.5 = 1 \times 0.5 = 0.5.$$

The probability of one boy before the first girl is

$$\Pr(Y = 1) = q^1 p = 0.5^1 \times 0.5 = 0.5 \times 0.5 = 0.25.$$

The probability of two boys before the first girl is

$$\Pr(Y = 2) = q^2 p = 0.5^2 \times 0.5 = 0.125.$$

and so on.

Expected Number of Failures before the First Success

For the geometric distribution, the expected (mean) number of failures before the first success is $E(Y) = (1 - p)/p$.

For the anti-depressant example, with $p = 0.6$, the mean number of failures before the first success is $E(Y) = (1 - p)/p = (1 - 0.6)/0.6 = 0.67$.

For the kidney-donor example, with $p = 0.1$, the mean number of failures before the first success is $E(Y) = (1 - 0.1)/0.1 = 9$.

For the alternative formulation, where X is the number of trials up to and including the first success, the expected value is $E(X) = 1/p$.

Moments and Cumulants

The expected value of a geometrically distributed random variable X is $1/p$ and the variance is $(1 - p)/p^2$:

$$E(X) = \frac{1}{p}, \qquad \operatorname{var}(X) = \frac{1-p}{p^2}.$$

Similarly, the expected value of the geometrically distributed random variable $Y = X - 1$ (where Y corresponds to the pmf listed in the right column) is $q/p = (1 - p)/p$, and its variance is $(1 - p)/p^2$:

$$E(Y) = \frac{1-p}{p}, \qquad \operatorname{var}(Y) = \frac{1-p}{p^2}.$$

Let $\mu = (1 - p)/p$ be the expected value of Y. Then the cumulants κ_n of the probability distribution of Y satisfy the recursion

$$\kappa_{n+1} = \mu(\mu+1)\frac{d\kappa_n}{d\mu}.$$

Outline of proof: That the expected value is $(1-p)/p$ can be shown in the following way. Let Y be as above. Then

$$\mathrm{E}(Y) = \sum_{k=0}^{\infty}(1-p)^k p\cdot k = p\sum_{k=0}^{\infty}(1-p)^k k = p(1-p)\sum_{k=0}^{\infty}(1-p)^{k-1}\cdot k = p(1-p)\left[\frac{d}{dp}\left(-\sum_{k=0}^{\infty}(1-p)^k\right)\right] = p(1-p)\frac{d}{dp}\left(-\frac{1}{p}\right) = \frac{1-p}{p}.$$

(The interchange of summation and differentiation is justified by the fact that convergent power series converge uniformly on compact subsets of the set of points where they converge.)

Parameter Estimation

For both variants of the geometric distribution, the parameter p can be estimated by equating the expected value with the sample mean. This is the method of moments, which in this case happens to yield maximum likelihood estimates of p.

Specifically, for the first variant let $k = k_1, \ldots, k_n$ be a sample where $k_i \geq 1$ for $i = 1, \ldots, n$. Then p can be estimated as

$$\hat{p} = \left(\frac{1}{n}\sum_{i=1}^{n}k_i\right)^{-1} = \frac{n}{\sum_{i=1}^{n}k_i}.$$

In Bayesian inference, the Beta distribution is the conjugate prior distribution for the parameter p. If this parameter is given a Beta(α, β) prior, then the posterior distribution is

$$p \sim \mathrm{Beta}\left(\alpha+n,\ \beta+\sum_{i=1}^{n}(k_i-1)\right).$$

The posterior mean E[p] approaches the maximum likelihood estimate $\hat{p}$ as α and β approach zero.

In the alternative case, let $k_1, \ldots, k_n$ be a sample where $k_i \geq 0$ for $i = 1, \ldots, n$. Then p can be estimated as

$$\hat{p} = \left(1+\frac{1}{n}\sum_{i=1}^{n}k_i\right)^{-1} = \frac{n}{\sum_{i=1}^{n}k_i+n}.$$

The posterior distribution of p given a Beta(α, β) prior is

$$p \sim \mathrm{Beta}\left(\alpha+n,\ \beta+\sum_{i=1}^{n}k_i\right).$$

Again the posterior mean E[p] approaches the maximum likelihood estimate $\hat{p}$ as α and β approach zero.

Other Properties

- The probability-generating functions of X and Y are, respectively,

$$G_X(s) = \frac{sp}{1-s(1-p)},$$

$$G_Y(s) = \frac{p}{1-s(1-p)}, \quad |s| < (1-p)^{-1}.$$

- Like its continuous analogue (the exponential distribution), the geometric distribution is memoryless. That means that if you intend to repeat an experiment until the first success, then, given that the first success has not yet occurred, the conditional probability distribution of the number of additional trials does not depend on how many failures have been observed. The die one throws or the coin one tosses does not have a "memory" of these failures. The geometric distribution is the only memoryless discrete distribution.
- Among all discrete probability distributions supported on {1, 2, 3, ... } with given expected value μ, the geometric distribution X with parameter $p = 1/\mu$ is the one with the largest entropy.
- The geometric distribution of the number Y of failures before the first success is infinitely divisible, i.e., for any positive integer n, there exist independent identically distributed random variables $Y_1, \ldots, Y_n$ whose sum has the same distribution that Y has. These will not be geometrically distributed unless $n = 1$; they follow a negative binomial distribution.
- The decimal digits of the geometrically distributed random variable Y are a sequence of independent (and *not* identically distributed) random variables. For example, the hundreds digit D has this probability distribution:

$$\Pr(D = d) = \frac{q^{100d}}{1 + q^{100} + q^{200} + \cdots + q^{900}},$$

where $q = 1 - p$, and similarly for the other digits, and, more generally, similarly for numeral systems with other bases than 10. When the base is 2, this shows that a geometrically distributed random variable can be written as a sum of independent random variables whose probability distributions are indecomposable.

- Golomb coding is the optimal prefix code for the geometric discrete distribution.
- The following recurrence relation holds:

$$\left\{\begin{array}{l} (p-1)\Pr(k) + \Pr(k+1) = 0, \\ \Pr(0) = p \end{array}\right\}$$

Related Distributions

- The geometric distribution Y is a special case of the negative binomial distribution, with $r = 1$. More generally, if $Y_1, \ldots, Y_r$ are independent geometrically distributed variables with parameter p, then the sum

$$Z = \sum_{m=1}^{r} Y_m$$

follows a negative binomial distribution with parameters r and p.

- The geometric distribution is a special case of discrete Compound Poisson distribution.
- If $Y_1, \ldots, Y_r$ are independent geometrically distributed variables (with possibly different success parameters p_m), then their minimum

$$W = \min_{m \in 1,\ldots,r} Y_m$$

is also geometrically distributed, with parameter $p = 1 - \prod_m (1 - p_m)$.

- Suppose $0 < r < 1$, and for $k = 1, 2, 3, \ldots$ the random variable X_k has a Poisson distribution with expected value r^k/k. Then

$$\sum_{k=1}^{\infty} k X_k$$

has a geometric distribution taking values in the set $\{0, 1, 2, \ldots\}$, with expected value $r/(1 - r)$.

- The exponential distribution is the continuous analogue of the geometric distribution. If X is an exponentially distributed random variable with parameter λ, then

$$y = \lfloor X \rfloor,$$

where $\lfloor \quad \rfloor$ is the floor (or greatest integer) function, is a geometrically distributed random variable with parameter $p = 1 - e^{-\lambda}$ (thus $\lambda = -\ln(1 - p)$) and taking values in the set $\{0, 1, 2, \ldots\}$. This can be used to generate geometrically distributed pseudorandom numbers by first generating exponentially distributed pseudorandom numbers from a uniform pseudorandom number generator: then $\langle \ln(U)/\ln(1-p) \rangle$ is geometrically distributed with parameter p, if U is uniformly distributed in $[0,1]$.

- If $p = 1/n$ and X is geometrically distributed with parameter p, then the distribution of X/n approaches an exponential distribution with expected value 1 as $n \to \infty$, since

$$P(X > a) = (1-p)^a = \left(1 - \frac{1}{n}\right)^{n\frac{1}{n}(a)}$$

$$= \left[\left(1 - \frac{1}{n}\right)^n\right]^{\frac{1}{n}(a)} \to [e^{-1}]^{\frac{1}{n}(a)} = e^{-\frac{1}{n}a} \text{ as } n \to \infty.$$

Computer Software for the Geometric Distribution

Geometric Distribution using R

The R function dgeom(k, prob) calculates the probability that there are k failures before the first success, where the argument "prob" is the probability of success on each trial. For example,

```
dgeom(0,0.6) = 0.6
```

```
dgeom(1,0.6) = 0.24
```

R uses the convention that k is the number of failures, so that the number of trials up to and including the first success is $k + 1$.

The following R code creates a graph of the geometric distribution from $Y = 0$ to 10, with $p = 0.6$.

```
Y=0:10
```

```
plot(Y, dgeom(Y,0.6), type="h", ylim=c(0,1), main="Geometric distribution for p=0.6", ylab="P(Y=Y)", xlab="Y=Number of failures before first success")
```

The geometric distribution, for the number of failures before the first success, is a special case of the negative binomial distribution, for the number of failures before s successes.

The Excel function `NEGBINOMDIST(number_f, number_s, probability_s)` calculates the probability of k = number_f failures before s = number_s successes where p = probability_s is the probability of success on each trial. For the geometric distribution, let number_s = 1 success.

For example,

```
=NEGBINOMDIST(0, 1, 0.6) = 0.6
```

```
=NEGBINOMDIST(1, 1, 0.6) = 0.24
```

Like R, Excel uses the convention that k is the number of failures, so that the number of trials up to and including the first success is k + 1.

The Geometric Distribution

The geometric distribution is also based on a sequence of independent Bernoulli trials. It represents the probability of the first success occurring in the xth trial. This means that,

In other words t he first $(x\text{-}1)$ independent trials result in a failure and the last trial is a success. Since all the trials are independent the probability would be,

The binomial distribution expresses the probability of x successes in a sequence of n independent Bernoulli trials.

$P(x) = q^{x-1}\, p \; x = 1, 2, 3.....$

The mean and variance of the geometric distribution are

$E[X] = p^{-1} \; V[X] = qp^{2}$

The probability mass function of a Geometric random variable with p =0.2 is shown in the figure below.

Return Period

The concept of return period is important when dealing with the natural hazards such as earth quakes, severe cyclones, floods etc.

Suppose X_i is the random variable representing the maximum earthquake magnitude that may occur near particular fault in a given year i. For many years I = 1, 2, 3,... if X_i s are the independent random variables, the probability that the time interval between occurrence of earthquakes of magnitude greater than certain value equals to n years is,

$$\begin{aligned} p(T=n) &= p(Y_1 < y)\,p(Y_2 < y)\ldots p(y_{n-1} < y)\,p(Y_n > y) \\ &= p(Y < y)^{n-1}\,p(Y > y) \end{aligned}$$

When dealing with the natural hazards the return period is defined as the mean time interval between exceedances of specific value which is equals to the reciprocal of the probability of exceedances.

$$E(\tilde{T}) = 1/p(Y > y)$$

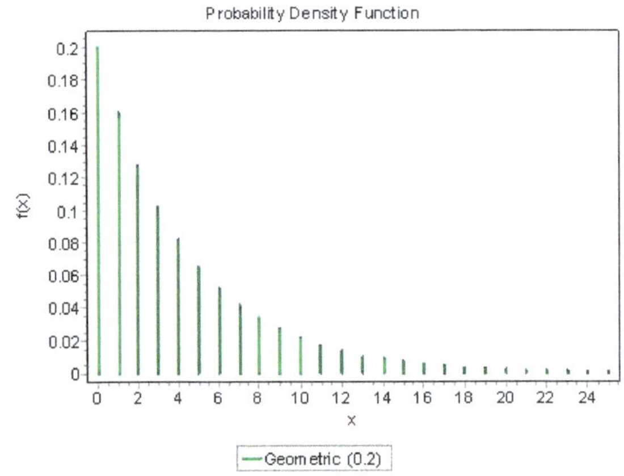

Probability mass function of a Geometric random variable with p 0.2

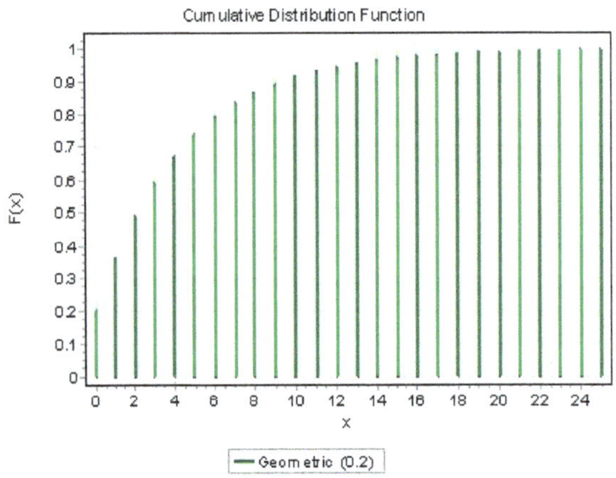

Cumulative distribution function of a Geometric random variable with p 0.2

Applications

In modeling the driver behavior, intersection turning movements and in speed studies this distribution is used. For e.g. if the probability of vehicle turning left at an intersection is 0.15 then the probability that the first left turning vehicle will be the fourth vehicle is:

$p(4) = (0.85)^3 (0.15) = 0.092$

Problem

For the data given in the concrete block problem, what is the probability that the fifth cube failing the test?

$p=0.95$

$q=0.05$

$p(X=5) = 0.954 \times 0.05 = 0.041$

Poisson Distribution

In probability theory and statistics, the Poisson distribution, named after French mathematician Siméon Denis Poisson, is a discrete probability distribution that expresses the probability of a given number of events occurring in a fixed interval of time and/or space if these events occur with a known average rate and independently of the time since the last event. The Poisson distribution can also be used for the number of events in other specified intervals such as distance, area or volume.

For instance, an individual keeping track of the amount of mail they receive each day may notice that they receive an average number of 4 letters per day. If receiving any particular piece of mail doesn't affect the arrival times of future pieces of mail, i.e., if pieces of mail from a wide range of sources arrive independently of one another, then a reasonable assumption is that the number of pieces of mail received per day obeys a Poisson distribution. Other examples that may follow a Poisson: the number of phone calls received by a call center per hour or the number of decay events per second from a radioactive source.

Introduction to the Poisson Distribution

The Poisson distribution is popular for modelling the number of times an event occurs in an interval of time or space.

Examples

The Poisson distribution may be useful to model events such as

- The number of meteors greater than 1 meter diameter that strike earth in a year

- The number of patients arriving in an emergency room between 10 and 11 pm

Assumptions: When is the Poisson Distribution an Appropriate Model?

The Poisson distribution is an appropriate model if the following assumptions are true.

- K is the number of times an event occurs in an interval and K can take values 0, 1, 2, ...
- The occurrence of one event does not affect the probability that a second event will occur. That is, events occur independently.
- The rate at which events occur is constant. The rate cannot be higher in some intervals and lower in other intervals.
- Two events cannot occur at exactly the same instant.
- The probability of an event in a small interval is proportional to the length of the interval.

If these conditions are true, then K is a Poisson random variable, and the distribution of K is a Poisson distribution.

Probability of Events for a Poisson Distribution

An event can occur 0, 1, 2, ... times in an interval. The average number of events in an interval is designated λ (lambda). Lambda is the event rate, also called the rate parameter. The probability of observing k events in an interval is given by the equation

$$P(k \text{ events in interval}) = e^{-\lambda} \frac{\lambda^k}{k!}$$

where

- λ is the average number of events per interval
- e is the number 2.71828... (Euler's number) the base of the natural logarithms
- k takes values 0, 1, 2, ...
- $k! = k \times (k-1) \times (k-2) \times \ldots \times 2 \times 1$ is the factorial of k.

This equation is the probability mass function (PMF) for a Poisson distribution.

Examples of Probability for Poisson Distributions

On a particular river, overflow floods occur once every 100 years on average. Calculate the probability of k = 0, 1, 2, 3, 4, 5, or 6 overflow floods in a 100-year interval, assuming the Poisson model is appropriate.

Because the average event rate is one overflow flood per 100 years, $\lambda = 1$

$$P(k \text{ overflow floods in 100 years}) = \frac{\lambda^k e^{-\lambda}}{k!} = \frac{1^k e^{-1}}{k!}$$

$$P(k = 0 \text{ overflow floods in 100 years}) = \frac{1^0 e^{-1}}{0!} = \frac{e^{-1}}{1} = 0.368$$

$$P(k = 1 \text{ overflow flood in 100 years}) = \frac{1^1 e^{-1}}{1!} = \frac{e^{-1}}{1} = 0.368$$

$$P(k = 2 \text{ overflow floods in 100 years}) = \frac{1^2 e^{-1}}{2!} = \frac{e^{-1}}{2} = 0.184$$

The table below gives the probability for 0 to 6 overflow floods in a 100-year period.

k	**P(*k* overflow floods in 100 years)**
0	0.368
1	0.368
2	0.184
3	0.061
4	0.015
5	0.003
6	0.0005

Ugarte and colleagues report that the average number of goals in a World Cup soccer match is approximately 2.5 and the Poisson model is appropriate.

Because the average event rate is 2.5 goals per match, $\lambda = 2.5$.

$$P(k \text{ goals in a match}) = \frac{2.5^k e^{-2.5}}{k!}$$

$$P(k = 0 \text{ goals in a match}) = \frac{2.5^0 e^{-2.5}}{0!} = \frac{e^{-2.5}}{1} = 0.082$$

$$P(k = 1 \text{ goal in a match}) = \frac{2.5^1 e^{-2.5}}{1!} = \frac{2.5e^{-2.5}}{1} = 0.205$$

$$P(k = 2 \text{ goals in a match}) = \frac{2.5^2 e^{-2.5}}{2!} = \frac{6.25e^{-2.5}}{2} = 0.257$$

The table below gives the probability for 0 to 7 goals in a match.

k	**P(*k* goals in a World Cup soccer match)**
0	0.082
1	0.205
2	0.257

3	0.213
4	0.133
5	0.067
6	0.028
7	0.010

Once in an Interval Events: The Special Case of λ = 1 and k = 0

Suppose that astronomers estimate that large meteors (above a certain size) hit the earth on average once every 100 years ($\lambda = 1$ event per 100 years), and that the number of meteor hits follows a Poisson distribution. What is the probability of $k = 0$ meteor hits in the next 100 years?

$$P(k = 0 \text{ meteors hit in next 100 years}) = \frac{1^0 e^{-1}}{0!} = \frac{1}{e} = 0.37$$

Under these assumptions, the probability that no large meteors hit the earth in the next 100 years is $p = 0.37$. The remaining $1 - 0.37 = 0.63$ is the probability of 1, 2, 3, or more large meteor hits in the next 100 years. In an example above, an overflow flood occurred once every 100 years ($\lambda = 1$). The probability of no overflow floods in 100 years was $p = 0.37$, by the same calculation.

In general, if an event occurs once per interval ($\lambda = 1$), and the events follow a Poisson distribution, then P($k = 0$ events in next interval) = 0.37.

As it happens, P(exactly one event in next interval) = 0.37, as shown in the table for overflow floods.

Examples that Violate the Poisson Assumptions

The number of students who arrive at the student union per minute will likely not follow a Poisson distribution, because the rate is not constant (low rate during class time, high rate between class times) and the arrivals of individual students are not independent (students tend to come in groups).

The number of magnitude 5 earthquakes per year in California may not follow a Poisson distribution if one large earthquake increases the probability of aftershocks of similar magnitude.

Among patients admitted to the intensive care unit of a hospital, the number of days that the patients spend in the ICU is not Poisson distributed because the number of days cannot be zero. The distribution may be modeled using a Zero-truncated Poisson distribution.

Count distributions in which the number of intervals with zero events is higher than predicted by a Poisson model may be modeled using a Zero-inflated model.

Poisson Regression and Negative Binomial Regression

Poisson regression and negative binomial regression are useful for analyses where the dependent (response) variable is the count (0, 1, 2, ...) of the number of events or occurrences in an interval.

History

The distribution was first introduced by Siméon Denis Poisson (1781–1840) and published, together with his probability theory, in 1837 in his work *Recherches sur la probabilité des jugements en matière criminelle et en matière civile* ("Research on the Probability of Judgments in Criminal and Civil Matters"). The work theorized about the number of wrongful convictions in a given country by focusing on certain random variables N that count, among other things, the number of discrete occurrences (sometimes called "events" or "arrivals") that take place during a time-interval of given length. The result had been given previously by Abraham de Moivre (1711) in *De Mensura Sortis seu; de Probabilitate Eventuum in Ludis a Casu Fortuito Pendentibus* in Philosophical Transactions of the Royal Society, p. 219. This makes it an example of Stigler's law and it has prompted some authors to argue that the Poisson distribution should bear the name of de Moivre.

A practical application of this distribution was made by Ladislaus Bortkiewicz in 1898 when he was given the task of investigating the number of soldiers in the Prussian army killed accidentally by horse kicks; this experiment introduced the Poisson distribution to the field of reliability engineering.

Definition

A discrete random variable X is said to have a Poisson distribution with parameter $\lambda > 0$, if, for $k = 0, 1, 2, \ldots$, the probability mass function of X is given by:

$$f(k;\lambda) = \Pr(X = k) = \frac{\lambda^k e^{-\lambda}}{k!},$$

where

- e is Euler's number ($e = 2.71828\ldots$)
- $k!$ is the factorial of k.

The positive real number λ is equal to the expected value of X and also to its variance

$$\lambda = \mathrm{E}(X) = \mathrm{Var}(X).$$

The Poisson distribution can be applied to systems with a large number of possible events, each of which is rare. How many such events will occur during a fixed time interval? Under the right circumstances, this is a random number with a Poisson distribution.

The conventional definition of the Poisson distribution contains two terms that can easily overflow on computers: λ^k and $k!$. The fraction of λ^k to $k!$ can also produce a rounding error which is very large compared to $e^{-\lambda}$, and therefore give an erroneous result. For numerical stability the Poisson probability mass function should therefore be evaluated as

$$f(k;\lambda) = \exp\{k \ln \lambda - \lambda - \ln \Gamma(k+1)\},$$

which is mathematically equivalent but numerically stable. The natural logarithm of the Gamma function can be obtained using the lgamma function in the C (programming language) standard library (C99 version), the gammaln function in MATLAB or SciPy, or the log_gamma function in Fortran 2008 and later.

Properties

Mean

- The expected value and variance of a Poisson-distributed random variable are both equal to λ.
- The coefficient of variation is $\lambda^{-1/2}$, while the index of dispersion is 1.
- The mean absolute deviation about the mean is

$$\mathrm{E}|X-\lambda| = 2\exp(-\lambda)\frac{\lambda^{\lfloor\lambda\rfloor+1}}{\lfloor\lambda\rfloor!}.$$

- The mode of a Poisson-distributed random variable with non-integer λ is equal to $\lfloor\lambda\rfloor$, which is the largest integer less than or equal to λ. This is also written as floor(λ). When λ is a positive integer, the modes are λ and $\lambda - 1$.
- All of the cumulants of the Poisson distribution are equal to the expected value λ. The nth factorial moment of the Poisson distribution is λ^n.
- The expected value of a Poisson process is sometimes decomposed into the product of *intensity* and *exposure* (or more generally expressed as the integral of an "intensity function" over time or space, sometimes described as "exposure").

Median

Bounds for the median (ν) of the distribution are known and are sharp:

$$\lambda - \ln 2 \le \nu < \lambda + \frac{1}{3}.$$

Higher Moments

- The higher moments m_k of the Poisson distribution about the origin are Touchard polynomials in λ:

$$m_k = \sum_{i=0}^{k} \lambda^i \left\{ {k \atop i} \right\},$$

where the {braces} denote Stirling numbers of the second kind. The coefficients of the polynomials have a combinatorial meaning. In fact, when the expected value of the Poisson distribution is 1, then Dobinski's formula says that the nth moment equals the number of partitions of a set of size n.

Sums of Poisson-distributed Random Variables

If $X_i \sim Pois(\lambda_i) i = 1,\ldots,n$ are independent, and $\lambda = \sum_{i=1}^{n} \lambda_i$, then $Y = \left(\sum_{i=1}^{n} X_i \right) \sim Pois(\lambda)$. A converse is Raikov's theorem, which says that if the sum of two independent random variables is Poisson-distributed, then so are each of those two independent random variables.

Other Properties

- The Poisson distributions are infinitely divisible probability distributions.
- The directed Kullback–Leibler divergence of Pois(λ_0) from Pois(λ) is given by

$$D_{\mathrm{KL}}(\lambda \mid \lambda_0) = \lambda_0 - \lambda + \lambda \log \frac{\lambda}{\lambda_0}.$$

- Bounds for the tail probabilities of a Poisson random variable $X \sim Pois(\lambda)$ can be derived using a Chernoff bound argument.

$$P(X \geq x) \leq \frac{e^{-\lambda}(e\lambda)^x}{x^x}, \text{ for } x > \lambda,,$$

$$P(X \leq x) \leq \frac{e^{-\lambda}(e\lambda)^x}{x^x}, \text{ for } x < \lambda.$$

- Recurrence relation:

$$\begin{Bmatrix} (k+1)f(k+1) - \lambda f(k) = 0, \\ f(0) = e^{-\lambda} \end{Bmatrix}$$

Poisson Races

Let $X \sim Poi(\lambda)$ and $Y \sim Poi(\mu)$ be independent random variables, with $\lambda < \mu$, then we have that

$$\frac{e^{-(\sqrt{\mu}-\sqrt{\lambda})^2}}{(\lambda+\mu)^2} - \frac{e^{-(\lambda+\mu)}}{2\sqrt{\lambda\mu}} - \frac{e^{-(\lambda+\mu)}}{4\lambda\mu} \leq P(X - Y \geq 0) \leq e^{-(\sqrt{\mu}-\sqrt{\lambda})^2}$$

The upper bound is proved using a standard Chernoff bound.

The lower bound can be proved by noting that $P(X - Y \geq 0 \mid X + Y = i)$ is the probability that $Z \geq \frac{i}{2}$, where $Z \sim Bin\left(i, \frac{\lambda}{\lambda+\mu}\right)$, which is bounded below by $\frac{1}{(i+1)^2}e^{\left(2D\left(0.5\|\frac{\lambda}{\lambda+\mu}\right)\right)}$, where D is relative entropy. Further noting that $X+Y \sim Poi(\lambda+\mu)$, and computing a lower bound on the unconditional probability gives the result.

Related Distributions

- If $X_1 \sim \text{Pois}(\lambda_1)$ and $X_2 \sim \text{Pois}(\lambda_2)$ are independent, then the difference $Y = X_1 - X_2$ follows a Skellam distribution.

- If $X_1 \sim \text{Pois}(\lambda_1)$ and $X_2 \sim \text{Pois}(\lambda_2)$ are independent, then the distribution of X_1 conditional on $X_1 + X_2$ is a binomial distribution.

  Specifically, given $X_1 + X_2 = k$, $X_1 \sim \text{Binom}(k, \lambda_1 / (\lambda_1 + \lambda_2))$.

  More generally, if $X_1, X_2, \ldots, X_n$ are independent Poisson random variables with parameters $\lambda_1, \lambda_2, \ldots, \lambda_n$ then

  given $\sum_{j=1}^{n} X_j = k$, $X_i \sim \text{Binom}\left(k, \frac{\lambda_i}{\sum_{j=1}^{n}\lambda_j}\right)$. In fact, $\{X_i\} \sim \text{Multinom}\left(k, \left\{\frac{\lambda_i}{\sum_{j=1}^{n}\lambda_j}\right\}\right)$.

- If $X \sim \text{Pois}(\lambda)$ and the distribution of Y, conditional on X = k, is a binomial distribution, $Y \mid (X=k) \sim \text{Binom}(k, p)$, then the distribution of Y follows a Poisson distribution $Y \sim \text{Pois}(\lambda \cdot p)$.. In fact, if $\{Y_i\}$, conditional on X = k, follows a multinomial distribution, $\{Y_i\} \mid (X=k) \sim \text{Multinom}\left(k, p_i\right)$, then each Y_i follows an independent Poisson distribution $Y_i \sim \text{Pois}(\lambda \cdot p_i), \rho(Y_i, Y_j) = 0$..

- The Poisson distribution can be derived as a limiting case to the binomial distribution as the number of trials goes to infinity and the expected number of successes remains fixed. Therefore, it can be used as an approximation of the binomial distribution if n is sufficiently large and p is sufficiently small. There is a rule of thumb stating that the Poisson distribution is a good approximation of the binomial distribution if n is at least 20 and p is smaller than or equal to 0.05, and an excellent approximation if $n \geq 100$ and $np \leq 10$.

$$F_{\text{Binomial}}(k; n, p) \approx F_{\text{Poisson}}(k; \lambda = np)$$

- The Poisson distribution is a special case of discrete compound Poisson distribution (or stuttering Poisson distribution) with only a parameter. The discrete compound Poisson distribution can be deduced from the limiting distribution of univariate multinomial distribution. It is also a special case of a compound Poisson distribution.

- For sufficiently large values of λ, (say λ>1000), the normal distribution with mean λ and variance λ (standard deviation $\sqrt{\lambda}$) is an excellent approximation to the Poisson distribution. If λ is greater than about 10, then the normal distribution is a good approximation if an appropriate continuity correction is performed, i.e., if $P(X \le x)$, where x is a non-negative integer, is replaced by $P(X \le x + 0.5)$.

$$F_{\text{Poisson}}(x;\lambda) \approx F_{\text{normal}}(x;\mu = \lambda, \sigma^2 = \lambda)$$

- Variance-stabilizing transformation: When a variable is Poisson distributed, its square root is approximately normally distributed with expected value of about $\sqrt{\lambda}$ and variance of about 1/4. Under this transformation, the convergence to normality (as λ increases) is far faster than the untransformed variable. Other, slightly more complicated, variance stabilizing transformations are available, one of which is Anscombe transform.

- If for every $t > 0$ the number of arrivals in the time interval $[0, t]$ follows the Poisson distribution with mean λt, then the sequence of inter-arrival times are independent and identically distributed exponential random variables having mean $1/\lambda$.

- The cumulative distribution functions of the Poisson and chi-squared distributions are related in the following ways:

$$F_{\text{Poisson}}(k;\lambda) = 1 - F_{\chi^2}(2\lambda; 2(k+1)) \qquad \text{integer } k,$$

and

$$\Pr(X = k) = F_{\chi^2}(2\lambda; 2(k+1)) - F_{\chi^2}(2\lambda; 2k).$$

Occurrence

Applications of the Poisson distribution can be found in many fields related to counting:

- Telecommunication example: telephone calls arriving in a system.
- Astronomy example: photons arriving at a telescope.
- Biology example: the number of mutations on a strand of DNA per unit length.
- Management example: customers arriving at a counter or call centre.
- Finance and insurance example: number of Losses/Claims occurring in a given period of Time.
- Earthquake seismology example: an asymptotic Poisson model of seismic risk for large earthquakes. (Lomnitz, 1994).
- Radioactivity example: number of decays in a given time interval in a radioactive sample.

The Poisson distribution arises in connection with Poisson processes. It applies to various phenomena of discrete properties (that is, those that may happen 0, 1, 2, 3, ... times during a given period of time or in a given area) whenever the probability of the phenomenon happening is constant in time or space. Examples of events that may be modelled as a Poisson distribution include:

- The number of soldiers killed by horse-kicks each year in each corps in the Prussian cavalry. This example was made famous by a book of Ladislaus Josephovich Bortkiewicz (1868–1931).
- The number of yeast cells used when brewing Guinness beer. This example was made famous by William Sealy Gosset (1876–1937).
- The number of phone calls arriving at a call centre within a minute. This example was made famous by A.K. Erlang (1878 – 1929).
- Internet traffic.
- The number of goals in sports involving two competing teams.
- The number of deaths per year in a given age group.
- The number of jumps in a stock price in a given time interval.
- Under an assumption of homogeneity, the number of times a web server is accessed per minute.
- The number of mutations in a given stretch of DNA after a certain amount of radiation.
- The proportion of cells that will be infected at a given multiplicity of infection.
- The arrival of photons on a pixel circuit at a given illumination and over a given time period.
- The targeting of V-1 flying bombs on London during World War II investigated by R. D. Clarke in 1946.

Gallagher in 1976 showed that the counts of prime numbers in short intervals obey a Poisson distribution provided a certain version of an unproved conjecture of Hardy and Littlewood is true.

Law of Rare Events

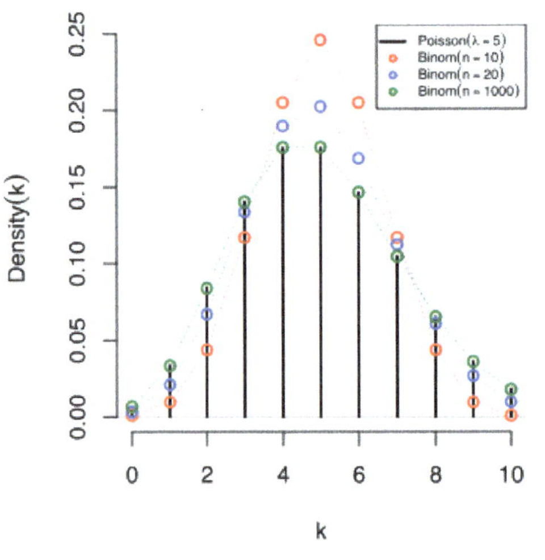

Comparison of the Poisson distribution (black lines) and the binomial distribution with n = 10 (red circles), n = 20 (blue circles), n = 1000 (green circles). All distributions have a mean of 5. The horizontal axis shows the number of events k. Notice that as n gets larger, the Poisson distribution becomes an increasingly better approximation for the binomial distribution with the same mean.

The rate of an event is related to the probability of an event occurring in some small subinterval (of time, space or otherwise). In the case of the Poisson distribution, one assumes that there exists a small enough subinterval for which the probability of an event occurring twice is "negligible". With this assumption one can derive the Poisson distribution from the Binomial one, given only the information of expected number of total events in the whole interval. Let this total number be λ. Divide the whole interval into n subintervals $I_1,\ldots,I_n$ of equal size, such that $n >$ λ (since we are interested in only very small portions of the interval this assumption is meaningful). This means that the expected number of events in an interval I_i for each i is equal to λ / n. Now we assume that the occurrence of an event in the whole interval can be seen as a Bernoulli trial, where the ith trial corresponds to looking whether an event happens at the subinterval I_i with probability λ / n. The expected number of total events in n such trials would be λ, the expected number of total events in the whole interval. Hence for each subdivision of the interval we have approximated the occurrence of the event as a Bernoulli process of the form $\mathrm{B}(n, \lambda / n)$. As we have noted before we want to consider only very small subintervals. Therefore, we take the limit as n goes to infinity. In this case the binomial distribution converges to what is known as the Poisson distribution by the Poisson limit theorem.

In several of the above examples—such as, the number of mutations in a given sequence of DNA—the events being counted are actually the outcomes of discrete trials, and would more precisely be modelled using the binomial distribution, that is

$$X \sim \mathrm{B}(n, p).$$

In such cases n is very large and p is very small (and so the expectation np is of intermediate magnitude). Then the distribution may be approximated by the less cumbersome Poisson distribution

$$X \sim \mathrm{Pois}(np).$$

This approximation is sometimes known as the law of rare events, since each of the n individual Bernoulli events rarely occurs. The name may be misleading because the total count of success events in a Poisson process need not be rare if the parameter np is not small. For example, the number of telephone calls to a busy switchboard in one hour follows a Poisson distribution with the events appearing frequent to the operator, but they are rare from the point of view of the average member of the population who is very unlikely to make a call to that switchboard in that hour.

The word law is sometimes used as a synonym of probability distribution, and *convergence in law* means *convergence in distribution*. Accordingly, the Poisson distribution is sometimes called the law of small numbers because it is the probability distribution of the number of occurrences of an event that happens rarely but has very many opportunities to happen. *The Law of Small Numbers* is a book by Ladislaus Bortkiewicz (Bortkevitch) about the Poisson distribution, published in 1898.

Poisson Point Process

The Poisson distribution arises as the number of points of a Poisson point process located in some finite region. More specifically, if D is some region space, for example Euclidean space R^d, for which $|D|$, the area, volume or, more generally, the Lebesgue measure of the region is finite, and if $N(D)$ denotes the number of points in D, then

$$P(N(D)=k)=\frac{(\lambda|D|)^k e^{-\lambda|D|}}{k!}.$$

Other Applications in Science

In a Poisson process, the number of observed occurrences fluctuates about its mean λ with a standard deviation $\sigma_k=\sqrt{\lambda}$. These fluctuations are denoted as Poisson noise or (particularly in electronics) as shot noise.

The correlation of the mean and standard deviation in counting independent discrete occurrences is useful scientifically. By monitoring how the fluctuations vary with the mean signal, one can estimate the contribution of a single occurrence, *even if that contribution is too small to be detected directly*. For example, the charge e on an electron can be estimated by correlating the magnitude of an electric current with its shot noise. If N electrons pass a point in a given time t on the average, the mean current is $I=eN/t$; since the current fluctuations should be of the order $\sigma_I=e\sqrt{N}/t$ (i.e., the standard deviation of the Poisson process), the charge e can be estimated from the ratio $t\sigma_I^2/I$.

An everyday example is the graininess that appears as photographs are enlarged; the graininess is due to Poisson fluctuations in the number of reduced silver grains, not to the individual grains themselves. By correlating the graininess with the degree of enlargement, one can estimate the contribution of an individual grain (which is otherwise too small to be seen unaided). Many other molecular applications of Poisson noise have been developed, e.g., estimating the number density of receptor molecules in a cell membrane.

$$\Pr(N_t=k)=f(k;\lambda t)=\frac{e^{-\lambda t}(\lambda t)^k}{k!}.$$

In Causal Set theory the discrete elements of spacetime follow a Poisson distribution in the volume.

Generating Poisson-distributed Random Variables

A simple algorithm to generate random Poisson-distributed numbers (pseudo-random number sampling) has been given by Knuth:

```
algorithm poisson random number (Knuth):

    init:

        Let L ← e^-λ, k ← 0 and p ← 1.

    do:

        k ← k + 1.

        Generate uniform random number u in [0,1] and let p ← p × u.
```

```
    while p > L.
    return k - 1.
```

While simple, the complexity is linear in the returned value *k*, which is λ on average. There are many other algorithms to overcome this. Some are given in Ahrens & Dieter. Also, for large values of λ, there may be numerical stability issues because of the term $e^{-\lambda}$. This could be solved by a slight change to allow λ to be added into the calculation gradually:

```
algorithm poisson random number (Junhao, based on Knuth):
    init:
        Let λLeft ← λ, k ← 0 and p ← 1.
    do:
        k ← k + 1.
        Generate uniform random number u in (0,1) and let p ← p × u.
        if p < e and λLeft > 0:
            if λLeft > STEP:
                p ← p × e^STEP
                λLeft ← λLeft - STEP
            else:
                p ← p × e^λLeft
                λLeft ← -1
    while p > 1.
    return k - 1.
```

The choice of STEP depends on the threshold of overflow. For double precision floating point format, the threshold is near e^{700}, so 500 shall be a safe *STEP*.

Other solutions for large values of λ include rejection sampling and using Gaussian approximation.

Inverse transform sampling is simple and efficient for small values of λ, and requires only one uniform random number *u* per sample. Cumulative probabilities are examined in turn until one exceeds *u*.

```
algorithm Poisson generator based upon the inversion by sequential search:
    init:
        Let x ← 0, p ← e^-λ, s ← p.
        Generate uniform random number u in [0,1].
    while u > s do:
```

```
        x ← x + 1.
        p ← p * λ / x.
        s ← s + p.
    return x.
```

"This algorithm ... requires expected time proportional to λ as λ→∞. For large λ, round-off errors proliferate, which provides us with another reason for avoiding large values of λ."

Parameter Estimation

Maximum Likelihood

Given a sample of n measured values $k_i = 0, 1, 2, \ldots$, for $i = 1, \ldots, n$, we wish to estimate the value of the parameter λ of the Poisson population from which the sample was drawn. The maximum likelihood estimate is

$$\hat{\lambda}_{\mathrm{MLE}} = \frac{1}{n}\sum_{i=1}^{n} k_i .$$

Since each observation has expectation λ so does this sample mean. Therefore, the maximum likelihood estimate is an unbiased estimator of λ. It is also an efficient estimator, i.e. its estimation variance achieves the Cramér–Rao lower bound (CRLB). Hence it is minimum-variance unbiased. Also it can be proved that the sum (and hence the sample mean as it is a one-to-one function of the sum) is a complete and sufficient statistic for λ.

To prove sufficiency we may use the factorization theorem. Consider partitioning the probability mass function of the joint Poisson distribution for the sample into two parts: one that depends solely on the sample $\mathbf{x}$ (called $h(\mathbf{x})$) and one that depends on the parameter λ and the sample $\mathbf{x}$ only through the function $T(\mathbf{x})$. Then $T(\mathbf{x})$ is a sufficient statistic for .

$$P(\mathbf{x}) = \prod_{i=1}^{n} \frac{\lambda^{x_i} e^{-\lambda}}{x_i!} = \frac{1}{\prod_{i=1}^{n} x_i!} \times \lambda^{\sum_{i=1}^{n} x_i} e^{-n\lambda}$$

Note that the first term, $h(\mathbf{x})$, depends only on $\mathbf{x}$. The second term, $g(T(\mathrm{x}) \mid \lambda)$, depends on the sample only through $T(\mathbf{x}) = \sum_{i=1}^{n} x_i$. Thus, $T(\mathbf{x})$ is sufficient.

To find the parameter λ that maximizes the probability function for the Poisson population, we can use the logarithm of the likelihood function:

$$\ell(\lambda) = \ln \prod_{i=1}^{n} f(k_i \mid \lambda) = \sum_{i=1}^{n} \ln\left(\frac{e^{-\lambda}\lambda^{k_i}}{k_i!}\right) = -n\lambda + \left(\sum_{i=1}^{n} k_i\right)\ln(\lambda) - \sum_{i=1}^{n} \ln(k_i!).$$

We take the derivative of ℓ with respect to λ and compare it to zero:

$$\frac{\mathrm{d}}{\mathrm{d}\lambda}\ell(\lambda) = 0 \Leftrightarrow -n + \left(\sum_{i=1}^{n} k_i\right)\frac{1}{\lambda} = 0.$$

Solving for λ gives a stationary point.

$$\lambda = \frac{\sum_{i=1}^{n} k_i}{n}$$

So λ is the average of the k_i values. Obtaining the sign of the second derivative of L at the stationary point will determine what kind of extreme value λ is.

$$\frac{\partial^2 \ell}{\partial \lambda^2} = -\lambda^{-2} \sum_{i=1}^{n} k_i$$

Evaluating the second derivative *at the stationary point* gives:

$$\frac{\partial^2 \ell}{\partial \lambda^2} = -\frac{n^2}{\sum_{i=1}^{n} k_i}$$

which is the negative of n times the reciprocal of the average of the k_i. This expression is negative when the average is positive. If this is satisfied, then the stationary point maximizes the probability function.

For completeness, a family of distributions is said to be complete if and only if $E(g(T)) = 0$ implies that $P_\lambda(g(T) = 0) = 1$ for all λ. If the individual X_i are iid $\mathrm{Po}(\lambda)$, then $T(\mathbf{x}) = \sum_{i=1}^{n} X_i \sim \mathrm{Po}(n\lambda)$.. Knowing the distribution we want to investigate, it is easy to see that the statistic is complete.

$$E(g(T)) = \sum_{t=0}^{\infty} g(t) \frac{(n\lambda)^t e^{-n\lambda}}{t!} = 0$$

For this equality to hold, it is obvious that $g(t)$ must be 0. This follows from the fact that none of the other terms will be 0 for all t in the sum and for all possible values of λ. Hence, $E(g(T)) = 0$ for all λ implies that $P_\lambda(g(T) = 0) = 1$, and the statistic has been shown to be complete.

Confidence Interval

The confidence interval for the mean of a Poisson distribution can be expressed using the relationship between the cumulative distribution functions of the Poisson and chi-squared distributions. The chi-squared distribution is itself closely related to the gamma distribution, and this leads to an alternative expression. Given an observation k from a Poisson distribution with mean μ, a confidence interval for μ with confidence level $1 - \alpha$ is

$$\tfrac{1}{2}\chi^2(\alpha/2; 2k) \le \mu \le \tfrac{1}{2}\chi^2(1-\alpha/2; 2k+2),$$

or equivalently,

$$F^{-1}(\alpha/2;k,1) \le \mu \le F^{-1}(1-\alpha/2;k+1,1),$$

where $\chi^2(p;n)$ is the quantile function (corresponding to a lower tail area p) of the chi-squared distribution with n degrees of freedom and $F^{-1}(p;n,1)$ is the quantile function of a Gamma distribution with shape parameter n and scale parameter 1. This interval is 'exact' in the sense that its coverage probability is never less than the nominal 1 – α.

When quantiles of the Gamma distribution are not available, an accurate approximation to this exact interval has been proposed (based on the Wilson–Hilferty transformation):

$$k\left(1-\frac{1}{9k}-\frac{z_{\alpha/2}}{3\sqrt{k}}\right)^3 \le \mu \le (k+1)\left(1-\frac{1}{9(k+1)}+\frac{z_{\alpha/2}}{3\sqrt{k+1}}\right)^3,$$

where $z_{\alpha/2}$ denotes the standard normal deviate with upper tail area α / 2.

For application of these formulae in the same context as above (given a sample of n measured values k_i each drawn from a Poisson distribution with mean λ), one would set

$$k = \sum_{i=1}^{n} k_i,$$

calculate an interval for $\mu = n\lambda$, and then derive the interval for λ.

Bayesian Inference

In Bayesian inference, the conjugate prior for the rate parameter λ of the Poisson distribution is the gamma distribution. Let

$$\lambda \sim \mathrm{Gamma}(\alpha, \beta)$$

denote that λ is distributed according to the gamma density g parameterized in terms of a shape parameter α and an inverse scale parameter β:

$$g(\lambda \mid \alpha, \beta) = \frac{\beta^\alpha}{\Gamma(\alpha)}\,\lambda^{\alpha-1}\,e^{-\beta\lambda} \qquad \text{for } \lambda > 0.$$

Then, given the same sample of n measured values k_i as before, and a prior of Gamma(α, β), the posterior distribution is

$$\lambda \sim \mathrm{Gamma}\left(\alpha + \sum_{i=1}^{n} k_i, \beta + n\right).$$

The posterior mean E[λ] approaches the maximum likelihood estimate $\hat{\lambda}_{\mathrm{MLE}}$ in the limit as $\alpha \to 0,\ \beta \to 0$ which follows immediately from the general expression of the mean of the gamma distribution.

The posterior predictive distribution for a single additional observation is a negative binomial distribution, sometimes called a Gamma–Poisson distribution.

Simultaneous Estimation of Multiple Poisson Means

Suppose $X_1, X_2, \ldots, X_p$ is a set of independent random variables from a set of p Poisson distributions, each with a parameter λ_i, $i = 1, \ldots, p$, and we would like to estimate these parameters. Then, Clevenson and Zidek show that under the normalized squared error loss $L(\lambda, \hat{\lambda}) = \sum_{i=1}^{p} \lambda_i^{-1} (\hat{\lambda}_i - \lambda_i)^2$, when $p > 1$, then, similar as in Stein's famous example for the Normal means, the MLE estimator $\hat{\lambda}_i = X_i$ is inadmissible.

In this case, a family of minimax estimators is given for any $0 < c \leq 2(p-1)$ and $b \geq (p - 2 + p^{-1})$ as

$$\hat{\lambda}_i = \left(1 - \frac{c}{b + \sum_{i=1}^{p} X_i} \right) X_i, \qquad i = 1, \ldots, p.$$

Bivariate Poisson Distribution

This distribution has been extended to the bivariate case. The generating function for this distribution is

$$g(u, v) = \exp[(\theta_1 - \theta_{12})(u - 1) + (\theta_2 - \theta_{12})(v - 1) + \theta_{12}(uv - 1)]$$

with

$$\theta_1, \theta_2 > \theta_{12} > 0$$

The marginal distributions are Poisson(θ_1) and Poisson(θ_2) and the correlation coefficient is limited to the range

$$0 \leq \rho \leq \min\left\{ \frac{\theta_1}{\theta_2}, \frac{\theta_2}{\theta_1} \right\}$$

A simple way to generate a bivariate Poisson distribution X_1, X_2 is to take three independent Poisson distributions Y_1, Y_2, Y_3 with means $\lambda_1, \lambda_2, \lambda_3$ and then set $X_1 = Y_1 + Y_3, X_2 = Y_2 + Y_3$. The probability function of the bivariate Poisson distribution is

$$\Pr(X_1 = k_1, X_2 = k_2) = \exp\left(-\lambda_1 - \lambda_2 - \lambda_3\right) \frac{\lambda_1^{k_1}}{k_1!} \frac{\lambda_2^{k_2}}{k_2!} \sum_{k=0}^{\min(k_1, k_2)} \binom{k_1}{k} \binom{k_2}{k} k! \left(\frac{\lambda_3}{\lambda_1 \lambda_2} \right)^k$$

Computer Software for the Poisson Distribution

Poisson Distribution using R

The R function dpois(x, lambda) calculates the probability that there are x events in an interval, where the argument "lambda" is the average number of events per interval.

For example,

```
dpois(x=0, lambda=1) = 0.3678794

dpois(x=1,lambda=2.5) = 0.2052125
```

The following R code creates a graph of the Poisson distribution from x= 0 to 8, with lambda=2.5.

```
x=0:8

px = dpois(x, lambda=2.5)

plot(x, px, type="h", xlab="Number of events k", ylab="Probability of k events", ylim=c(0,0.5), pty="s", main="Poisson distribution \n Probability of events for lambda = 2.5")
```

Poisson Distribution Using Excel

The Excel function POISSON(x, mean, cumulative) calculates the probability of x events where mean is lambda, the average number of events per interval. The argument cumulative specifies the cumulative distribution.

For example,

```
=POISSON(0, 1, FALSE) = 0.3678794

=POISSON(1, 2.5, FALSE) = 0.2052125
```

Poisson Distribution using Python (SciPy)

The function `scipy.stats.distributions.poisson.pmf(x, poissonLambda)` calculates the probability that there are x events in an interval, where the argument "poissonLambda" is the average number of events per interval.

Poisson Distribution using Mathematica

Mathematica supports the univariate Poisson distribution as PoissonDistribution[λ], and the bivariate Poisson distribution as MultivariatePoissonDistribution[θ_{12},{ $\theta_1 - \theta_{12}$, $\theta_2 - \theta_{12}$ }], including computation of probabilities and expectation, sampling, parameter estimation and hypothesis testing.

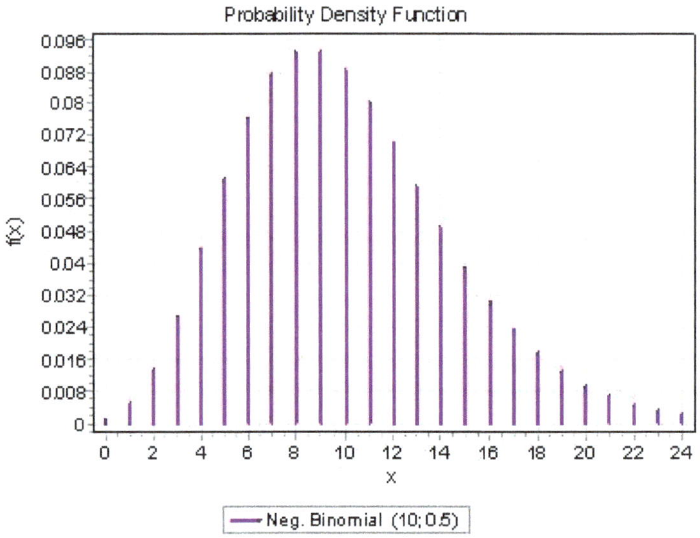

Probability mass function of a Negative binomial random variable with k=10 and p=0.5

The Negative Binomial or Pascal Distribution

The Pascal distribution measures the probability of the kth success occurring in the xth trial of a Bernoulli process.

In other words there must be (k-1) successes in the first (x-1) trials and the last trial must be a success.

According to the binomial distribution the probability of having k-1 successes in x-1 trials is

$$p(k-1)=\frac{(x-1)!}{(k-1)!(x-k)!}p^{k-1}q^{x-k}$$

Since the last trial results in a success the associated probability is p. Then the probability of thek^{th} success in x^{th} trial would be the product of the above two probabilities and it results in the following form,

$$P(x)=\frac{(x-1)!}{(k-1)!(x-k)!}p^{k}q^{x-k} \qquad For=k,k+1\ldots$$

Mean and the variance of the Pascal distribution are,

$$E[X] = kp^{-1} \qquad V[X]= kqp^{2}$$

The probability mass function of a Negative binomial variable with k = 10 and p= 0.5 is shown in the figure.

The geometric distribution is a special case of Pascal distribution when k=1. If the probability of vehicle turning left at an intersection is 0.15 then the probability that the eighth vehicle will be the third to turn left is

$$p(18)=\frac{7!*(0.15)^{3}*(0.85)^{\wedge 5}}{2!*7!}=0.031$$

The Poisson Distribution

This distribution has found wide applications in the field of Civil engineering. As n becomes large, while the value of np is a finite constant, the Binomial distribution approaches the Poisson distribution. Poisson distribution is normally valid when n is sufficiently large ($n>>50$) and p is small ($p<<0.1$).

$$P(r)=\frac{(np)^{r}e^{-np}}{r!}$$

Where, n may be interpreted as the number of time or space units into which a given time or space interval is divided. p is the probability of occurrence of an event (i.e. the probability of success in a Bernoulli trial) in a given time or space unit. r is the number of successes in n trials.

$$= \frac{m^r e^{-m}}{r!}$$

m = avg. no. of times the event occurs (avg. number of vehicles arrived in a given time interval or the avg. number of accidents occurred in a given space interval)

Development of the Poisson Distribution from the Binomial

The binomial distribution with c = np:

$$p(x) = \frac{n!}{x!(n-x)!} p^x (1-p)^{n-x} \qquad x = 0, 1, 2, \ldots n$$

Now,

$$np = c \Rightarrow p = \frac{c}{n} \Rightarrow 1 - p = 1 - \frac{c}{n} \Rightarrow 1 - p \frac{n-c}{n}$$

Replacing the terms involving p by corresponding terms involving c, we obtain:

$$p(x) = \frac{n(n-1)(n-2)\ldots(n-x+1)}{x!} \left[\frac{c}{n}\right]^x \left[\frac{n-c}{n}\right]^{n-x}$$

$$= \frac{c^x}{x!}\left[(1)\left(1-\frac{1}{n}\right)\left(1-\frac{2}{n}\right)\ldots\left(1-\frac{x-1}{n}\right)\right]\left(1-\frac{c}{n}\right)^n\left(1-\frac{c}{n}\right)^{-x}$$

Letting n → ∞ and p → 0 in such a way that np = c remains fixed, the terms:

$$\left(1-\frac{1}{n}\right), \left(1-\frac{2}{n}\right)\ldots\left(1-\frac{x-1}{n}\right) and \left(1-\frac{c}{n}\right)^{-x}_{\to 1}$$

It is known that

$$\left(1-\frac{c}{n}\right)^n \to e^c \, as \, n \to \infty$$

Thus, the limiting form of the previous equation becomes the Poisson distribution:

$$p(x) = \left(\frac{c^x}{x!}\right).e^{-c}$$

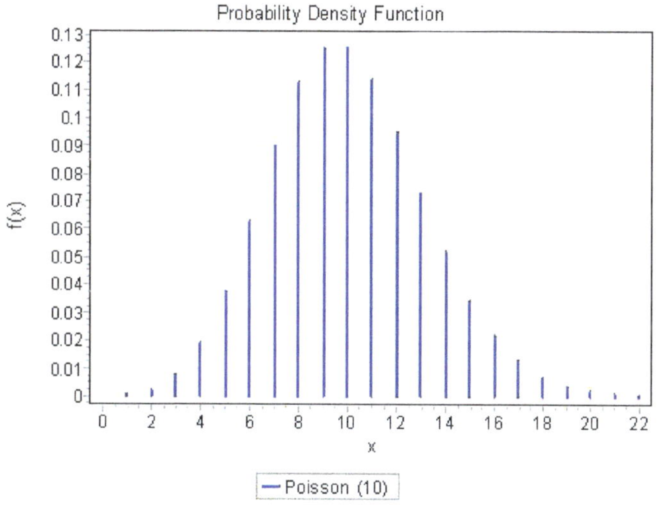

Probability mass function of a Poisson random variable with m=10

Mean and Variance of the Poisson Distribution

$$E(X)=\sum_{X=0}^{\infty}\frac{xe^{-c}c^{x}}{x!}=\sum_{X=1}^{\infty}\frac{e^{-c}c^{x}}{(x\quad 1)!}$$

$$=ce\left[\quad+\frac{c}{1!}+\frac{c}{2!}+\quad\right]$$

$$ce^{\;c}\,e^{c}$$

Similarly,

$$E(X^2)=\sum_{X=0}^{\infty}\frac{x^2e^{-c}c^{x}}{x!}=c^2+c$$

So that

$$V(X)=E(X^2)-[E(X)]^2=c$$

Problem

In the concrete block problem, what is the probability that the 10th test is the second failure?

p=0.95

q= 0.05

It is nothing but the probability of 2nd failure in 10th test, in a Bernoulli sequence of trials;

$$p(X=10)=\frac{(10-1)!}{(2-1)!(10-2)!}0.05^2*0.95^{10-2}=0.149$$

Applications of Poisson Distribution

Poisson distribution has several applications in civil engineering. In earth quake engineering, in a specific time interval the probability of occurrence of an earth quake at a particular fault follows Poisson distribution. Occurrence of cyclones in a particular time period follows Poisson distribution.

Example 1:

From the past experience it is known that on an average every two years 3 cyclones hit the coastal area of Andhra Pradesh and Orissa states. If it is assumed that the cyclone hitting the coastal areas follows Poisson distribution then what is the probability of two cyclones crossing the coastal area of Andhra Pradesh and Orissa in the next two years?

$$p(X=2)=\frac{e^{-3}3^2}{2}=0.224$$

Example 2:

The most widely used situation is the arrival pattern of vehicles. In this case m becomes the average number of vehicles per any stated time interval. For this case we may write:

$$P(r)=\frac{(\lambda t)^r e^{-\lambda t}}{r!}$$

λ = the average vehicular arrival rate

Table: Sample calculations on finding the theoretical probabilities associated to a Poisson random variable

Vehicular arrivals (10secs)	Observed frequency (f_i)	Observed probability	Total no. of vehicles	Total time	Theoretical probability
0	11	0.11	0	110	0.1353
1	28	0.28	28	280	0.2706
2	30	0.30	60	300	0.2706
3	18	0.18	54	180	0.1804
4	8	0.08	32	80	0.0932
5	4	0.04	20	40	0.0361
6	1	0.01	6	10	0.0120
>=7	0	0.0	0	0	0.0048

- = total no. of vehicles arrived/ total time of observation

  = 200/1000 = 0.2 vehicles/sec

$$P(r) = \frac{(2)^r e^{-2}}{r!}$$

Sample calculations involved in calculating the theoretical probabilities associated to a Poisson random variable are shown in the table above. Observed and the theoretical probability mass functions are also shown in the figure below .

Other application of Poisson distribution is with the parking of vehicles.

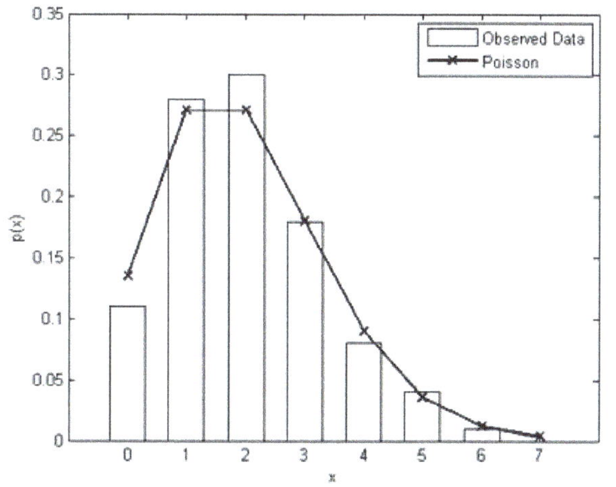

Modeling of vehicular arrivals in 10 sec intervals using the Poisson distribution

The Uniform Distribution

The uniform distribution describes the experiments that have a finite number of *N* equi-probable outcomes. The casting of a fair die and the tossing of a fair coin are two examples for this distribution.

$$P(x) = P[X = x] = \frac{1}{N}$$
$$Mean = E[X] = \sum_{\forall x} xp(x_i)$$

Since X can take any of the probable N outcomes ranging between 1 to N. Hence,

$$\sum_{\forall x} x * \frac{1}{N} = \frac{1}{N}\frac{N(N+1)}{2} = \frac{(N+1)}{2}$$
$$Variance = V[X]$$
$$Variance = \sum_i (x_i - \mu)^2 P(x) = \sum_i \left(x_i - \frac{(N+1)}{2} \right)^2 \frac{1}{N} = \frac{(N^2+1)}{12}$$

Probability mass function and the cumulative distribution function associated to a uniform random variable are shown in the figures below.

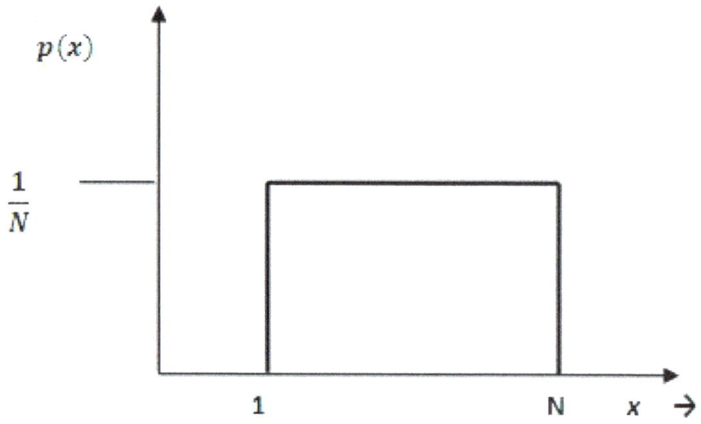

Probability mass function of a discrete uniform random variable

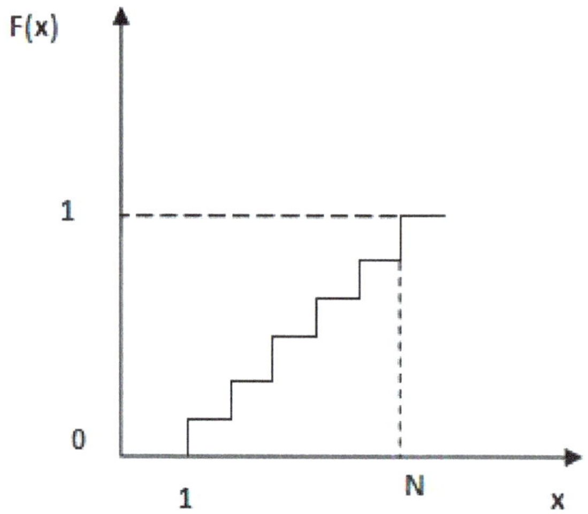

Cumulative distribution function of a discrete uniform random variable

Continuous Distributions

In most of the cases the output from the natural processes as well as the experimental outcomes related to the civil engineering takes up infinite number of values. Consider the measurement of the headways between vehicles crossing a road section. Conceivably, the headway between any two vehicles can vary from zero (in case of multilane roads and when they enter simultaneously) to infinity (following the last vehicle never to arrive at the road section). Assuming that the headways cannot be measured with absolute precision, the probability of obtaining exactly any particular value is zero.

The Uniform Distribution

The uniform distribution is defined as

$$f(x) = \frac{1}{(b-a)} \; for\; a \le x \le b$$

It consists of a horizontal line over the range in such a way that the area under the curve is equals to 1. The CDF of the uniform random variable is as follows;

$$F(x) = p(X \leq x) = 0 \; for \; x < a$$
$$= \frac{x-a}{b-a} \; for \; a \leq x < b$$
$$= 1 \; for \; x \geq b$$

The mean and the variance of the uniform probability density function are;

$$E[X] = \frac{a+b}{2} \qquad V[X] = \frac{(b-a)^2}{12}$$

Mean and the Variance

$$E[X] = \int_a^b \frac{x\,dx}{b-a} = \frac{1}{2(b-a)} x^2 \Big|_a^b = \frac{(b+a)}{2}$$
$$V[X] = \int_a^b \frac{x^2\,dx}{b-a} - \left[\frac{(b+a)}{2}\right]^2 = \frac{(b+a)^2}{12}$$

The Negative Exponential Distribution

The negative exponential distribution bears a special relationship with the discrete Poisson distribution. When the occurrence of an event follows the Poisson distribution the interval between the occurrences is distributed according to the negative exponential distribution. If the vehicles arrive at an intersection according to the Poisson distribution, the inter arrival times are exponentially distributed.

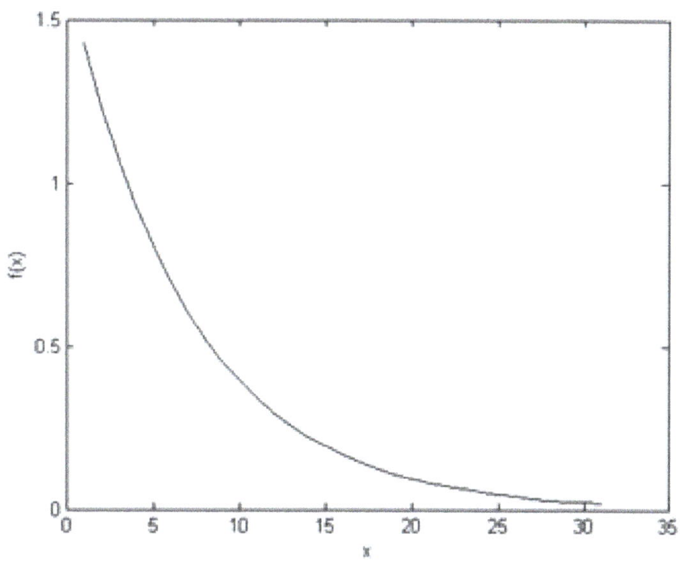

Probability density function for the negative exponential distribution

The probability density function for the negative exponential distribution is:

$$f(x) = \lambda e^{-\lambda x}$$

The mean and the variances are,

$$E[X] = \frac{1}{\lambda} \qquad V[X] = \frac{1}{\lambda^2}$$

The probability density function of an exponential random variable is shown in the Figure.

Relation of the Exponential Distribution to the Poisson Distribution

As stated earlier, if the number of occurrences follows a Poisson distribution, then the time between occurrences would follow an exponential distribution

The exponential distribution has the following probability density function:

$$\begin{aligned} f(x) &= \lambda e^{-\lambda x} & x \geq 0 \\ &= 0 & \textit{otherwise} \end{aligned}$$

Poisson distribution is developed from the Poisson postulates and the Poisson process, fixing time at some value t, and then developing the distribution of the number of occurrences in the interval [o, t]. Noting this random variable as X the distribution is:

$$\begin{aligned} p(x) &= e^{-\lambda t}\frac{(\lambda t)^x}{x!} & x = 0,12,\ldots \\ &= 0 & \textit{otherwise} \end{aligned}$$

The probability of zero occurrences in [o, t], p(o) will be,

$$p(0) = e^{-\lambda t}$$

The same can be written as the probability of occurrence of the first event after time't'

$$p(0) = p(T > t) = e^{-\lambda t} \qquad t \geq 0$$

The cumulative distribution function, i.e. the probability of occurrence of an event before time t, the above equation can be adjusted as given below,

$$F(t) = P(T \leq t) = 1 - e^{-\lambda t} \qquad t \geq 0$$

It is known that the probability density function f (t) is the derivative of the cumulative distribution function, F (T). Therefore,

$$\begin{aligned} f(t) &= \lambda e^{-\lambda t} & t \geq 0 \\ &= 0 & \textit{otherwise} \end{aligned}$$

This is the probability density function of the exponential distribution as stated in the beginning.

Figure (a) shows the observed headway data for traffic flows up to 14 vehicles/minute. Figure (b) shows the observed frequencies along with the theoretical exponential distribution with parameters obtained from the field data.

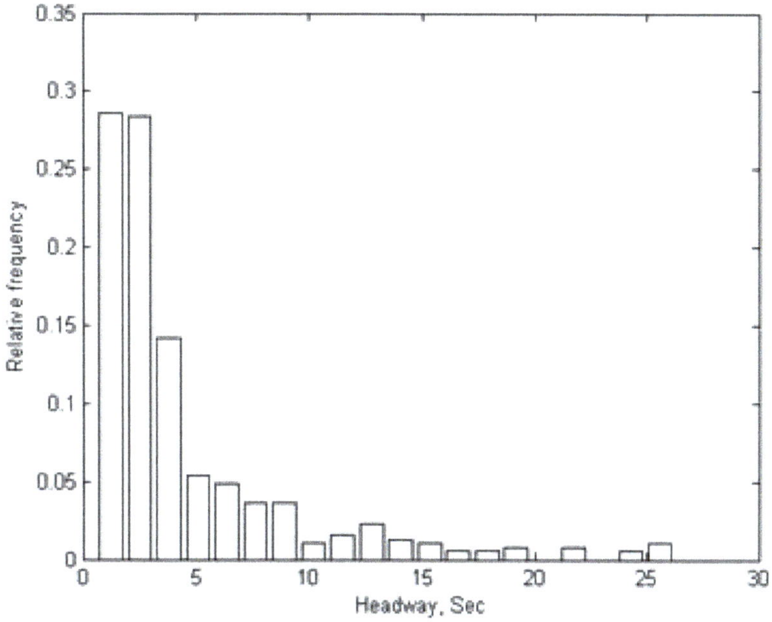

(a) Fitting of headway data collected in traffic flows up to14 vehicles/min to the exponential distribution ($\lambda = 0.22$)

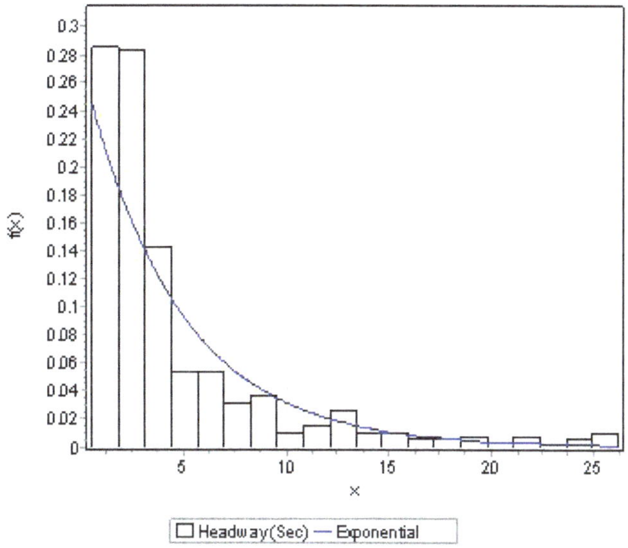

(b) Fitting of exponential distribution to the headway data

Mean and the Variance of the Exponential Distribution

Exponential random variable takes the positive values only, i.e. time of occurrence of an event will always be positive.

$$E(X) = \int_0^{\infty} x\lambda e^{-\lambda x}\, dx = -xe^{-\lambda x} \Big|_0^{\infty} + \int_0^{\infty} e^{-\lambda x}\, dx$$

$$= \frac{1}{\lambda}$$

$$V(X) = \int_0^{\infty} x\lambda e^{-\lambda x}\, dx - \left(\frac{1}{\lambda}\right)^2$$

$$= \left[-x^2\lambda e^{-\lambda x}\Big|_0^{\infty} + 2\int_0^{\infty} xe^{-\lambda x}\, dx\right] - \left(\frac{1}{\lambda}\right)^2$$

$$= \frac{1}{\lambda^2}$$

This mean is the reciprocal of the mean of the Poisson distribution. For example if the average number of vehicles arriving at a road section is the mean of a poison distribution, the inverse of that, i.e. the average time gap between the vehicular arrivals, would be the mean of the exponential distribution.

For example, if the cars arrive at the parking garage at a rate of 90 vehicles/hr, according to the Poisson distribution the mean headway E[x] = 1/λ =1/1.5 min/veh.

The probability that the mean headway is less than or equal to 45 sec. is

$$P(X \leq 0.75) = F(0.75) = 1 - E^{-(1.5*0.75)} = 0.675$$

Cumulative distribution function of the exponential random variable.

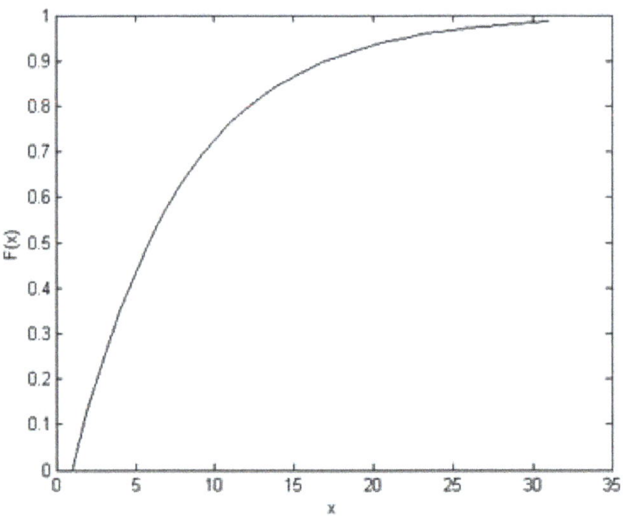

Cumulative distribution function

Problem

Cars arrive at an intersection at an average rate of 6 per minute. What is the probability that at least one car arrives in the next 30 secs?

It can be assumed that the arrival of vehicles at an intersection follows Poisson distribution. Based on this it can be said that the inter arrival times follow negative exponential distribution.

Average rate of arrival = 6 vehicles/min.

Average arrival time = 10 sec.

The density function for negative exponential distribution is,

$$f(x) = \lambda e^{-\lambda x}$$
$$F(x) = p(X \le x) = 1 - e^{-\lambda x}$$
$$p(X \ge x) = e^{-\lambda x}$$

Probability of at least one vehicle arriving at the intersection in the next 30 seconds is,

= 1 – probability of no vehicles arriving at the intersection

= 1 – probability of headway greater than or equal to 30 seconds

$$= 1 - p(X \ge 30) = 1 - e^{-\lambda x}$$
$$= 1 - e^{-\frac{30}{10}}$$
$$= 1 - 0.0497 = .95$$

Shifted Negative Exponential Distribution

When the exponential distribution is applied in the headway modeling, particularly for the low-flow conditions where the vehicles arrive randomly, one may not see the headways that are less than 1 sec. In this scenario, to fit the exponential distribution to the headway data it is necessary to use a shift parameter as shown below.

The figure below shows the shifted negative exponential distribution (shift parameter is 0.6 sec) along with the observed frequencies.

$$f(t) = \lambda e^{-\lambda(t-\alpha)}$$

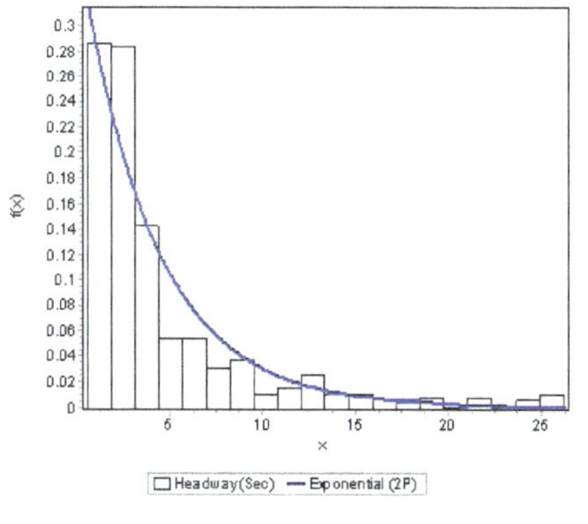

Modeling of headway data using shifted-negative exponential distribution for the low traffic conditions (λ = 0.25 and α = 0.6)

Central Limit Theorem

If a random variable Y is the sum of n independent random variables that satisfy the following general conditions, for sufficiently large n, the variable Y is approximately normally distributed.

General conditions

1. The individual normal random variable, X_i, contribute a negligible amount to the variance of the sum of all the n random variables
2. It is also unlikely that a single random variable contributing significantly to the sum

If $X_1, X_2, X_3, \ldots, X_n$ are the set of n independent random variables with i^{th} random variable having mean μ_i and variance σ^2_i, the resulting variable from the summation of all the n random variables is approximately normally distributed with mean $\sum_{i=1}^{n} \mu_i$ and standard deviation $\sqrt{\sum_{i=1}^{n} \sigma_i^2}$. The following standardized variable,

$Z = \frac{Y - \sum_{i=1}^{n} \mu_i}{\sqrt{\sum_{i=1}^{n} \sigma_i^2}}$ is an approximate standard normal distribution as n approaches infinity.

A linear transformation, $Y = a + bX$, of a random variable X with mean μ and variance σ^2 results in a normal variable Y with mean $a+b\mu$ and the variance $b^2\sigma^2$.

The Normal Distribution

The most important distribution of a continuous random variable is the normal distribution. This has got several applications in Civil engineering. One of the illustrative applications in traffic engineering is with the distribution of the speed data. This distribution and its properties are very useful in dealing with the samples. It can also be used for approximating the other types of the probability distributions.

A continuous random variable X is said to follow the normal distribution if its pdf is

$$f(x) = \frac{1}{\sigma\sqrt{2\pi}} e^{\frac{-(x-\mu)^2}{2\sigma^2}}, -\infty < x < \infty$$

The cumulative distribution function for the normal variable is given below;

$$F(x) = P(X \le x) = \frac{1}{\sigma\sqrt{2\pi}} \int_{-\infty}^{x} e^{\frac{-(u-u)^2}{2\sigma^2}} du, -\infty < x < \infty$$

It is difficult to evaluate the above integral hence the random variable is transformed into a standard normal variable and the resulting CDF is used in obtaining the corresponding probabilities. The transformation is as follows;

$$Z = \frac{(x-\mu)}{\sigma}$$

The resulting pdf , with mean zero and variance one, is;

$$f(z)=\frac{1}{\sqrt{2\pi}}e^{\frac{-z^2}{2}},\ -\infty<z<\infty$$

The CDF of the standard normal variable is;

$$F(z)=P(Z\le z)=\frac{1}{\sqrt{2\pi}}\int_{-\infty}^{z}e^{\frac{u^2}{2}}du,-\infty<x<\infty$$

Standard tables are available to get the F (z) values for different z values.

Properties of the Normal Distribution

Following are the important properties of the Normal Distribution:

i. $\int_{-\infty}^{\infty}f(x)dx=1$

ii. $f(x)\ge 0 \quad for\ all\ x$

iii. $\lim_{x\to\infty}f(x)=0\ and\ \lim_{x\to-\infty}f(x)=0$

iv. $f[(x+\mu)]=f[-(x-\mu)]$ *The density is symmetric about* μ

v. The maximum value of f occurs at x=μ

vi. The points of inflection of f are x=μ±σ

Proof for property (i):

A random variable X is said to have a normal distribution with mean μ (-∞ < μ< ∞) and variance σ^2 (σ^2 > o) if it has the density function:

$$f(x)=\frac{1}{\sigma\sqrt{2\pi}}e^{-\frac{1}{2}\left(\frac{x-\mu}{\sigma}\right)^2}\quad -\infty<x<\infty$$

Integrating and replacing $\frac{x-\mu}{\sigma}$ with y:

$$I=\frac{1}{\sqrt{2\pi}}\int_{-\infty}^{\infty}e^{-\frac{1}{2}y^2}dy$$

Defining a second normally distributed variable, Z

$$I^2=\frac{1}{\sqrt{2\pi}}\int_{-\infty}^{\infty}e^{-\frac{1}{2}y^2}dy\frac{1}{\sqrt{2\pi}}\int_{-\infty}^{\infty}e^{-\frac{1}{2}z^2}dz$$

$$=\frac{1}{\sqrt{2\pi}}\int_{-\infty}^{\infty}\int_{-\infty}^{\infty}e^{(y^2+z^2)}dy\,dz$$

Changing to polar coordinates with:

$$y = r\sin\theta$$
$$z = r\cos\theta$$

The integral becomes:

$$I^2 = \frac{1}{\sqrt{2\pi}}\int_0^\infty\int_0^{2\pi} re^{-\frac{1}{2}r^2}\,d\theta\,dr$$
$$= \int_0^\infty re^{-\frac{1}{2}r^2}\,dr$$
$$= 1$$

As $I^2=1$, $I=1$ since f must be positive everywhere.

Mean and the Variance of Normal Distribution

Normal random variable takes any value between -∞ to +∞. Expectation for this variable is given below;

$$E(X) = \int_{-\infty}^{\infty} x * \frac{1}{\sigma\sqrt{2\pi}} e^{-\frac{1}{2}\left(\frac{x-\mu}{\sigma}\right)^2} dx$$

By replacing (x-μ)/σ with z we get:

$$E(X) = \int_{-\infty}^{\infty} \frac{1}{\sqrt{2\pi}}(\mu+\sigma z)e^{-\frac{z^2}{2}}\,dz$$

$$\mu = \int_{-\infty}^{\infty} \frac{1}{\sqrt{2\pi}} e^{-\frac{z^2}{2}}\,dz + \sigma\int_{-\infty}^{\infty}\frac{1}{\sqrt{2\pi}} ze^{-\frac{z^2}{2}dz}$$

Of the two terms, value of the first term becomes one as the integrand has μ=0 and $\sigma^2=1$. The second term reduces to zero as:

$$\int_{-\infty}^{\infty}\frac{1}{\sqrt{2\pi}}ze^{-\frac{z^2}{2}}\,dz = -\frac{1}{\sqrt{2\pi}}ze^{-\frac{z^2}{2}}\Bigg|_{-\infty}^{\infty} = 0$$

And thus,

$$E(X) = \mu[1] + \sigma[0]$$
$$= \mu$$

To find the variance,

$$V(X) = E\left[(X-\mu)\right]^2$$

$$= \int_{-\infty}^{\infty} (x-\mu)^2 \frac{1}{\sigma\sqrt{2\pi}} e^{-\frac{1}{2}\left(\frac{x-\mu}{\sigma}\right)^2} dx$$

By replacing (x-μ)/σ with z we get:

$$V(X) = \int_{-\infty} \sigma^2 Z^2 \frac{1}{\sqrt{2\pi}} e^{-\frac{z^2}{2}} dz$$

$$= \sigma^2 \left[\int_{-\infty}^{\infty} \frac{z^2}{\sqrt{2\pi}} e^{-\frac{z^2}{2}} dz \right]$$

$$= \sigma^2 \left[\frac{z^2}{\sqrt{2\pi}} e^{-\frac{z^2}{2}} \Big|_{-\infty}^{\infty} + \int_{-\infty}^{\infty} \frac{1}{\sqrt{2\pi}} e^{-\frac{z^2}{2}} dz \right]$$

$$= \sigma^2 [0+1]$$

$$= \sigma^2$$

Figure (a) shows the probability density function of a normal random variable. Figure (b) shows the probability density function of the corresponding standard normal variable. Figure (c) shows the cumulative distribution function of a normal variable with mean 60 and variance 25.

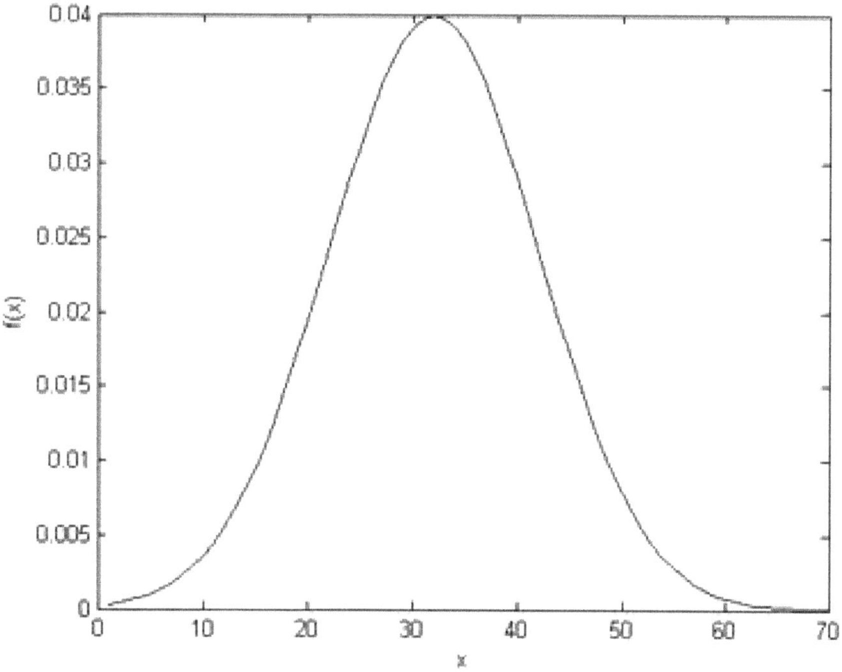

(a) Probability density function of a normal random variable

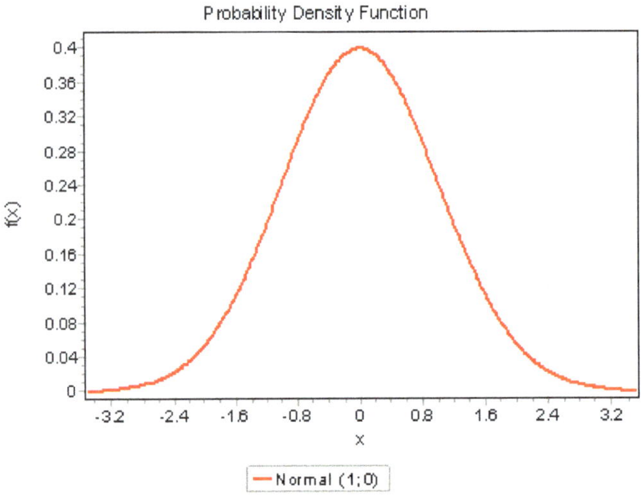

(b) Probability density function of a standard normal variable

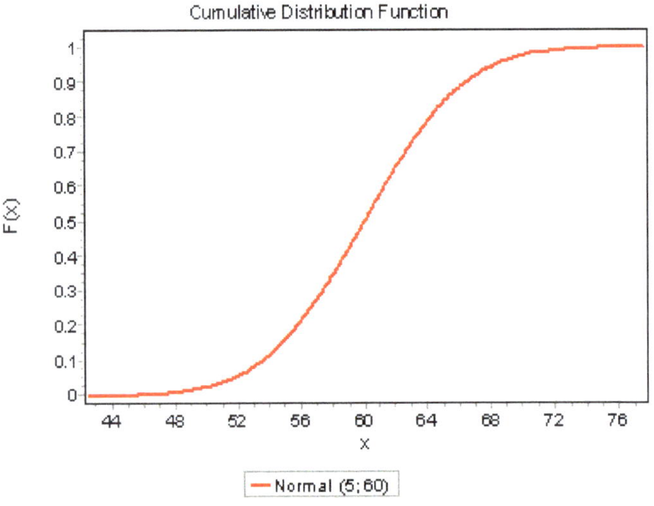

(c) Cumulative distribution function of a normal random variable with mean 60 and variance 25

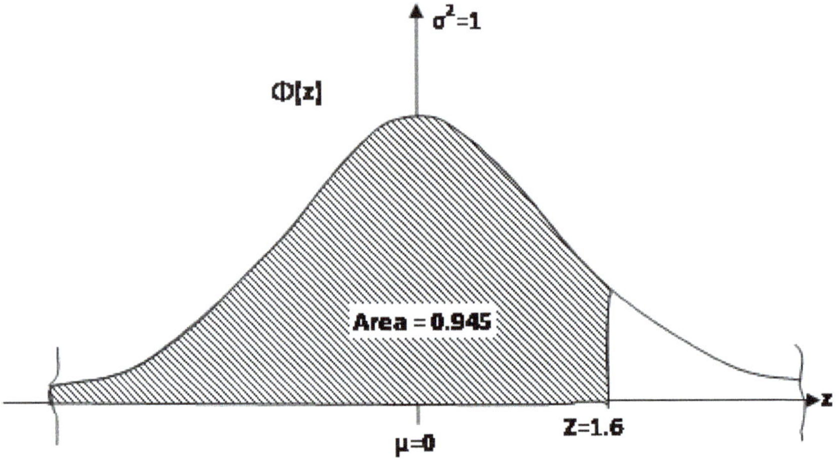

(d) Standard Normal pdf showing the area under the curve for z=1.6

For example, if the spot speeds at a location are normally distributed with mean of 51.7 kmph and a standard deviation of 8.3 kmph. Then the probability of speeds exceeding 65 kmph can be obtained as discussed below:

The standard normal variable, $Z = \dfrac{65-51.7}{8.3} = 1.6$

Area under the curve for z = 1.6 is 0.9452. This area gives the probability of vehicles having speed less than or equals to 65 kmph. To get the probability that the speed is greater than 65kmph this area must be deducted from 1.

The probability of having speed greater than 65 kmph is 1 - 0.9452 = 0.0458.

Figure (e) shows the frequency distribution of the speed data and the corresponding theoretical normal distribution.

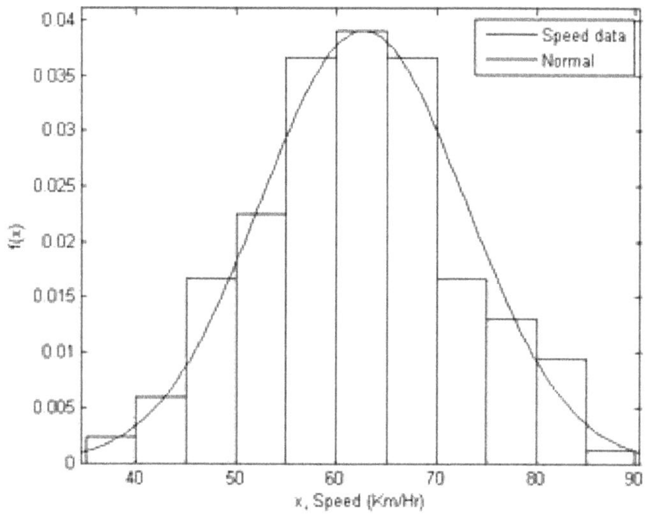

(e) Modeling the free flow speed of a particular vehicle type using the normal pdf

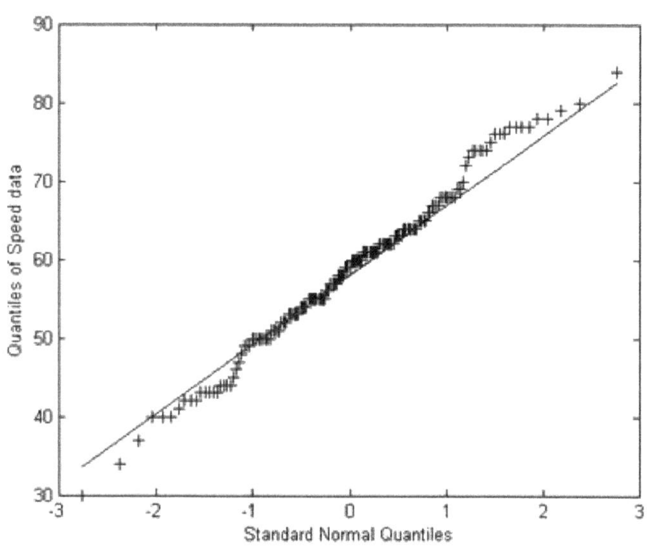

(f) Normal quantile or the probability plot

Figure (f) shows the normal quantile plot for the speed data. This plot is known as the probability plot and the procedure to draw this is discussed when dealing with Gumbel distribution. Similarly, Figure (g) shows the probability plot of the concrete cubes' compressive strength data.

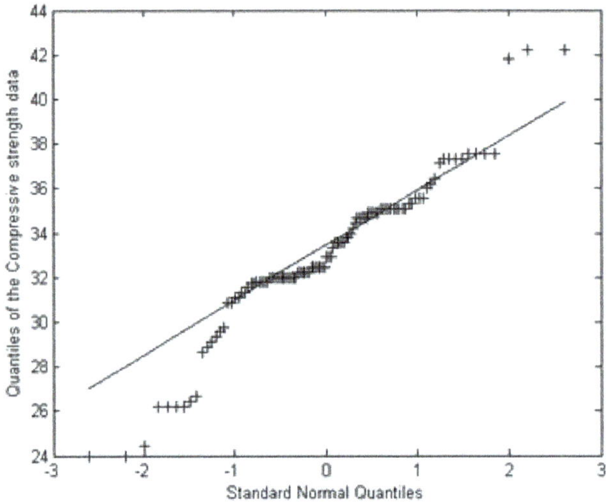

(g) Normal quantile or the probability plot of the concrete cube's compressive strength

Lognormal Distribution

It shows the outcome of single tailed experiment. Any real valued function whose logarithm is normally distributed, the function follows log normal distribution.

$$f(x)=\frac{1}{x\sigma\sqrt{\sqrt{2\pi}}}\exp\left[\frac{-1}{2}\{(In\,x-\mu)/\sigma\}^2\right]$$

Evaluation of probabilities associated with the lognormal distribution is simple due to its relation with the normal distribution. If the probability of a lognormal variable falling between a and b is to be found, the standard normal tables can be utilized in the following manner;

$$p(a<X<b)=\int_a^b\frac{1}{x\sigma\sqrt{2\pi}}\exp\left[\frac{-1}{2}\{(In\,x-\mu)/\sigma\}^2\right]dx$$

By replacing {In x-μ)/σ} with z, the above integral results in the following form;

$$p(a<X<b)=\frac{1}{\sqrt{2\pi}}\int_{(In\,a-\mu)/\sigma}^{(In\,b-\mu)/\sigma}\exp\left[\frac{-z^2}{2}\right]dz$$

The above integral can be easily evaluated using the areas under the standard normal curves corresponding to the two limits obtained from the standard normal tables. This distribution is useful in analyzing the processes where the outcomes are always positive. For example, the rain fall intensity, air traffic, suspended particulate matter in the ambient atmosphere, and the material strength characteristics are the processes where the outcome is always greater than or equals to zero. The figure below shows the probability density functions of the lognormal random variable corresponding to various parameters. The following figure shows the frequency distribution of the runoff data and the corresponding theoretical lognormal distribution. From this figure it can be seen that the lognormal distribution is closely approximating the runoff data.

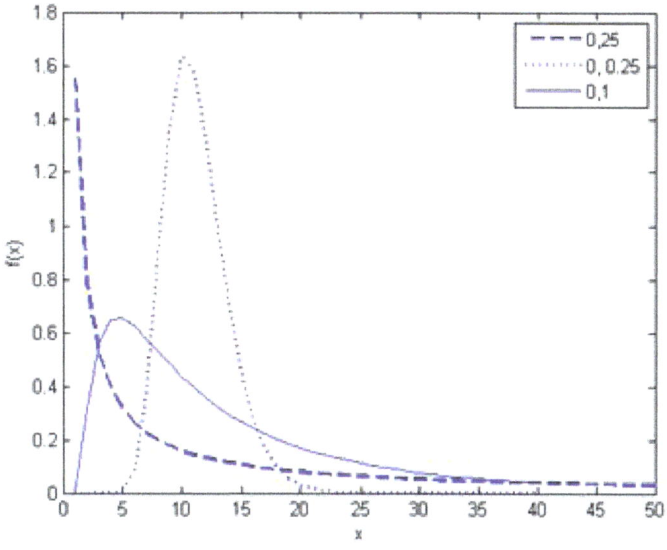

Lognormal distributions resulting from various combinations of means and variances of logX

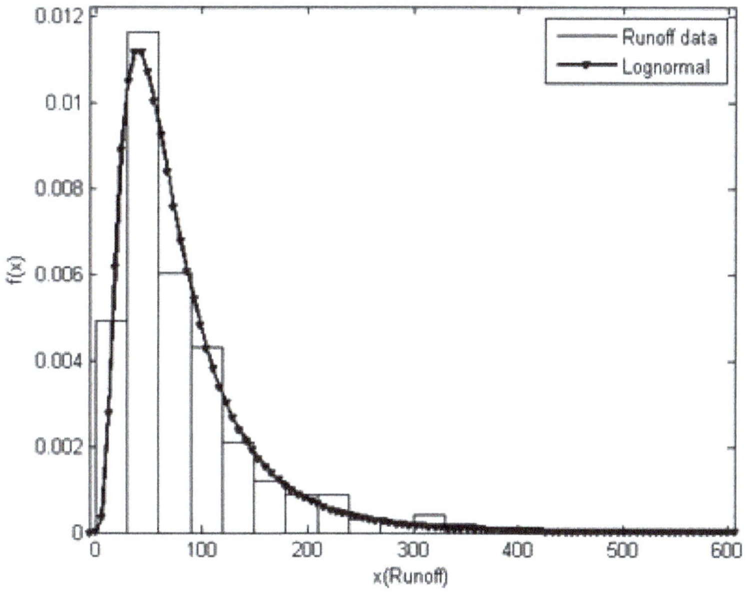

Runoff data modeling using lognormal distribution

Mean and the Variance of the Lognormal Distribution

A random variable X with range space $R_X = \{x: 0 < x < \infty\}$ is considered where Y=ln X is normally distributed with mean μ_y and variance σ_y^2.

Mean

$$E(X) = \mu_X = e^{\mu_Y + \frac{1}{2}\sigma_Y^2}$$

Variance

$$V(X) = \sigma_X^2 = e^{2\mu_Y + \sigma_y^2}\left(e^{\sigma_y^2} - 1\right)$$

Similar to the additive reproductive property of the normal distribution, this has the multiplicative reproductive property. If any two random variables X_1 and X_2 follow lognormal distribution with means μ_1, μ_2 and the variances σ_1^2, σ_2^2 respectively, the resulting variable from the multiplication of these two variables follow lognormal distribution with parameters $\mu_1 + \mu_2$ and $\sigma_1^2 + \sigma_2^2$.

Some other Important Continuous Distributions

The Gamma Distribution

Gamma distribution has several applications in civil engineering. One of the applications is in rainfall modeling. Variants of the Gamma distribution are widely used in the field of traffic engineering. This distribution is related to the Poisson process and the relationship is explained below. The figure a shows the probability density functions of the Gamma distribution with different shape parameters. Figure b shows the frequency distribution of the headway data and the corresponding probability density function of the Gamma random variable.

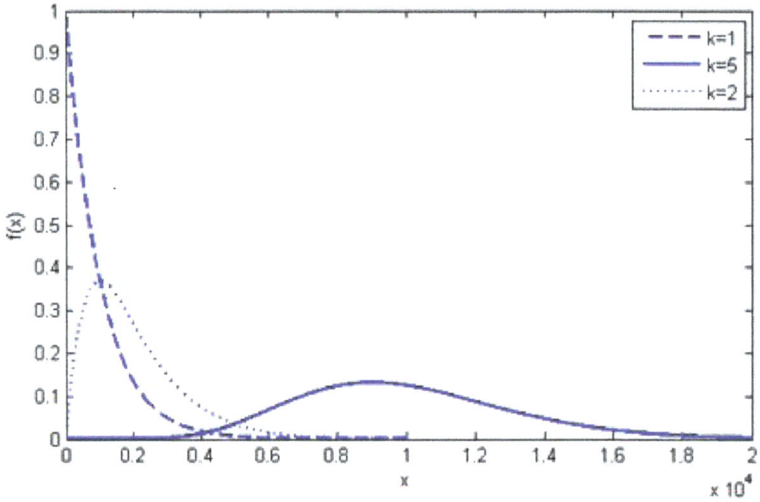

(a) PDF of Gamma distribution for various shape parameters

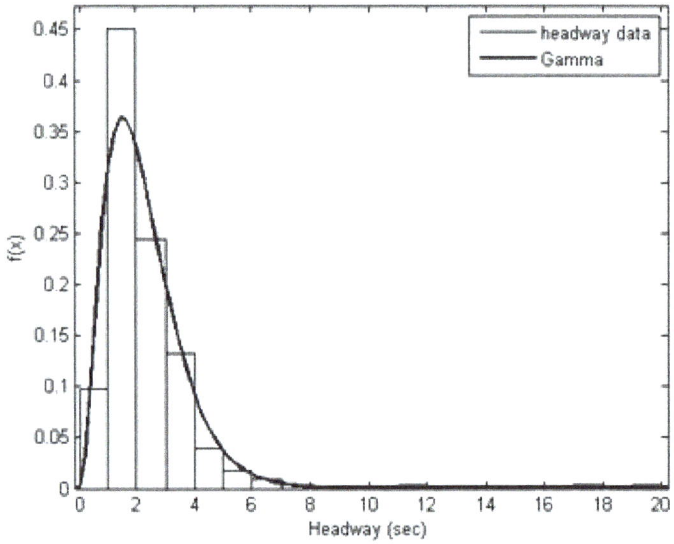

(b) Modeling of headway data using Gamma distribution

Relationship between the Gamma and the Exponential Distribution Functions

If the random variable X is the sum of r independent, exponentially distributed random variables, each with parameter λ, then X follows Gamma distribution with parameters r and λ.

$$X = X_1 + X_2 + X_3 + \ldots + X_r$$

The resulting variable X has the following probability density function;

$$f(x) = \frac{\lambda}{\Gamma(k)} (\lambda x)^{k-1} e^{-\lambda x} \; for\, 0 < x < \infty$$

Where, k and λ are the parameters of the Gamma distribution which are always greater than zero. λ is the scale parameter and k is the shape parameter.

Gamma function is defined as follows

$\Gamma(k) = \int_0^\infty x^{k-1} e^{-x} dx$ for k > o

When k is the positive integer, then

$\Gamma(k) = (k-1)!$

Mean and the Variance of Gamma Distribution

When r is an integer and also when $X = X_1 + X_2 + X_3 + \quad + X$

$$E(X) = \sum_{j=1}^{r} E(X_j)$$

$$= r.\frac{1}{\lambda}$$

$$= \frac{r}{\lambda}$$

And

$$V(X) = \sum_{j=1}^{r} V(X_j)$$

$$= r.\frac{1}{\lambda^2}$$

$$= \frac{r}{\lambda^2}$$

Applications of Gamma Distribution in Civil Engineering

From the above discussion it is clear that the Gamma distribution is related the exponential distribution (related to the Poisson process). If the events are occurring according to the Poisson distribution the time till the occurrence of the first event is described by the exponential distribution. The time till the occurrence of the second event is described by the Gamma distribution with r =2 (summation of the time intervals between two consecutive events).

For example it is known that under free flow conditions vehicular arrival pattern follows Poisson distribution. If the mean arrival rate is 5vehicles/minute, the headways between consecutive vehicular arrivals follow exponential distribution with the following pdf;

$$f(t)=\frac{1}{5}*e^{-5t}$$

The time gap between every two vehicles follows the Gamma distribution with the following pdf;

$$f(t)=\frac{1}{5}*(\frac{1}{5}*t)e^{-5t}$$

The time at which r^{th} vehicle arrives at the measurement location follows the gamma distribution with the following pdf;

$$f(t)=\frac{1}{5}*\frac{\left(\frac{1}{5}*t\right)^{r-1}}{(r-1)!}*e^{-t/5}$$

$$f(t)=\lambda*\frac{(\lambda t)^{r-1}}{(r)}*e^{-\lambda t}$$

Headway data following the exponential distribution is shown in Figure a. Similarly the summation of the first two, three and five headways fitted to the Gamma distribution are also shown in Figures b to d. The parameters of the theoretical Gamma distributions used in fitting the data prove the above relationship.

The above relationship is true if the r value is integer. When r takes integer values only the probability density function is also known as Erlang distribution.

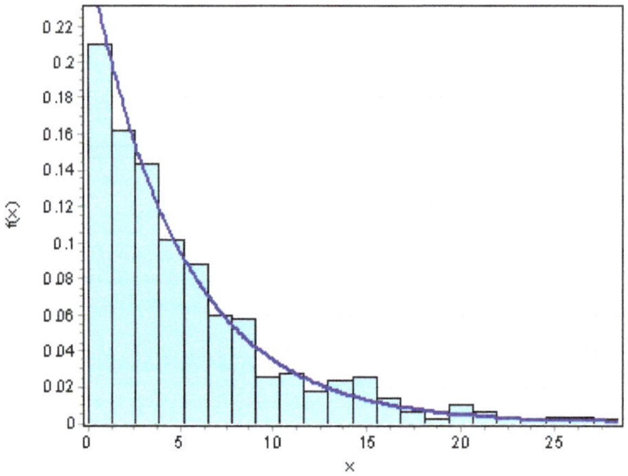

(a) Headway data (with mean arrival rate of 5Veh/minute) fitted to the exponential distribution

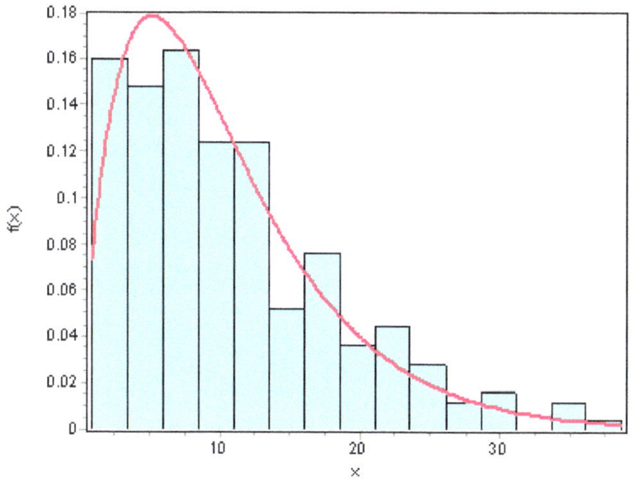

(b) Summation of two consecutive headway data fitted to the Gamma distribution with parameters r=2, λ=0.19

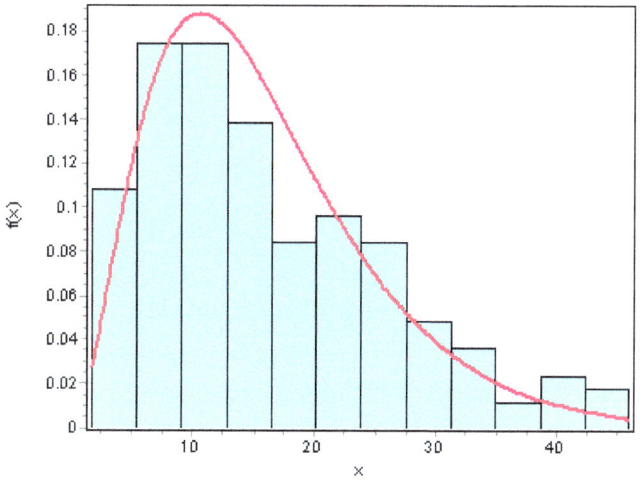

(c) Summation of three consecutive headway data fitted to the Gamma distribution with parameters r=3, λ=0.19

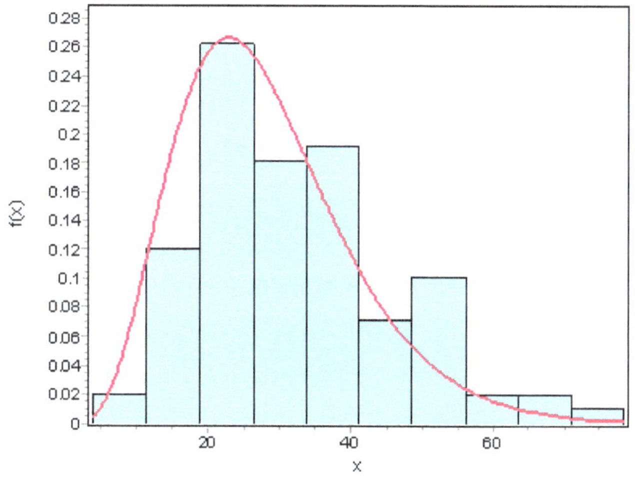

(d) Summation of three consecutive headway data fitted to the Gamma distribution with parameters r=5.3, λ=0.19

Problem

The life of E. coli (in days) has a gamma distribution with shape parameter α = 8 and scale parameter β = 15. What is the probability that thi s bacteria will live between 2 to 3 months?

$$f(x) = \frac{\lambda}{\Gamma(k)}(\lambda x)^{k-1} \; for\, 0 < x < \infty \left(\beta = \frac{1}{\lambda} and\, \alpha = k \right)$$

Answer:

$$F(x) = 1 - e^{-\lambda t}\left[\sum_{i=0}^{k-1} \frac{(\lambda t)^k}{k!}\right]$$

Since the shape parameter is integer the approximate form for the cumulative distribution function is,

Probability of bacteria living between 2 to 3 months is;

$$\begin{aligned} p(60 \le X \le 90) &= F(90) - F(60) \\ &= 0.256 - 0.05 \\ &= 0.206 \end{aligned}$$

Erlang Distribution

If the shape parameter *k* takes only the integer values, the Gamma distribution results in Erlang distribution.

$$f(t) = \frac{\lambda}{(k-1)!}[\lambda t]^{k-1} e^{-\lambda t}$$

Erlang distribution is useful in modeling the headways under high traffic flow conditions. In the above pdf, k is the shape parameter and is the scale parameter. Mean and the variances for this distribution are the same as that of the Gamma distribution.

If the mean and the standard deviations of the data are known, the shape and the scale parameters can be obtained as follows;

$$k = Nearest\, \text{int}\, eger\, to \left(\frac{mean}{s\, \tan dard\, deviation} \right)^2$$

$$\lambda = \frac{k}{mean}$$

Cumulative Distribution Function

For the integer values of the shape parameter k, the cumulative distribution function can be written as follows;

$$f(t)=1-e^{-\lambda t}\left[\sum_{i=0}^{k-1}\frac{(\lambda t)^k}{k!}\right]$$

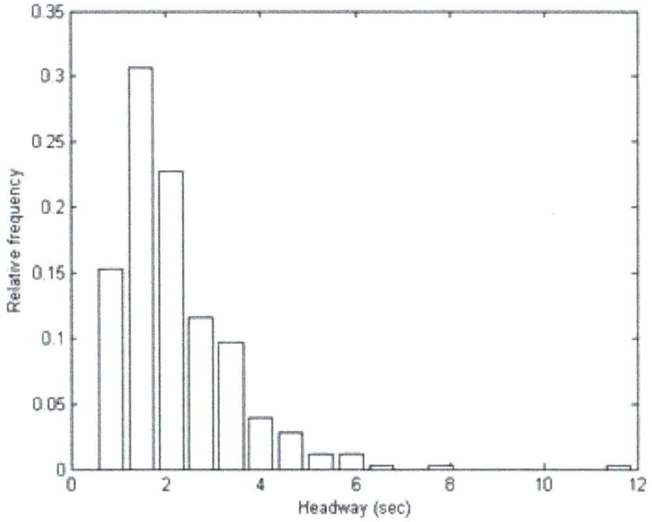

(a) Relative frequency histogram of the headway data observed under high traffic flow conditions on a single lane road

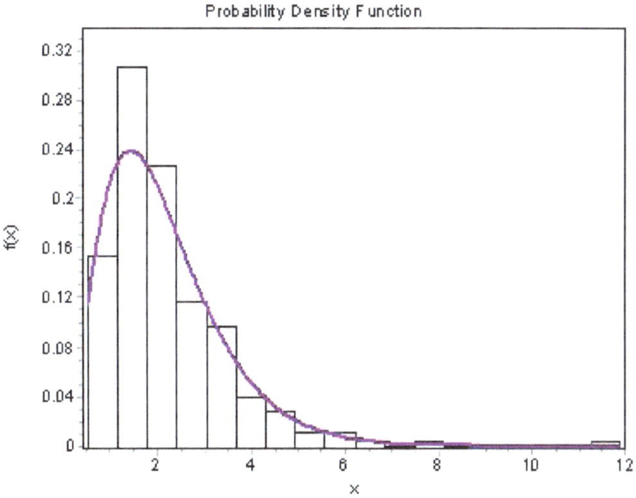

(b) Observed and Erlang densities of resulting from the headway data

Relative frequency histogram of the headway data is shown in Figure (a). Fitting the headway data observed in high traffic conditions has been shown in Figure (b). The shape and the scale parameters of the Erlang distribution are 3 and 1.376, respectively. The theoretical Erlang density function with the above calculated parameters along with the observed densities, are shown in Figure a.

Pearson Type III Distribution

Pearson type III distribution is also known as the three-parameter Gamma distribution. In addition to the shape and scale parameters, the shift parameter is used in this distribution. A random variable is said to follow the Pearson type III distribution if the probability density function is as given below;

$$f(t)=\frac{\lambda^{k}}{\Gamma(k)}(t-c)^{k-1}e^{-\lambda(t-c)} \quad for\, c<t<\infty$$

Where $\Gamma(k)$ is known as the Gamma function which is defined as follows;

$$\Gamma(k)=\int_{0}^{\infty}x^{k-1}e^{-x}dx$$

When this distribution is applied for modeling the headways observed on the multilane roads the shift parameter is useful to restrict the possibility of having very small headways.

This implies that the Pearson random variable takes values that are greater than c.

Extreme Value Distributions

Extreme value distributions are quite useful in civil engineering where the extreme conditions, which occur rarely, are the critical design elements. This is particularly true when considering the natural hazards such as the maximum intensity of the earth quake in the life span of a building, maximum flood levels in the life span of a bridge. These distributions are also useful in travel behavior modeling where the decisions related to the mode choice are presumed to follow extreme value distribution. Gumbel distribution, which is also known as the type 1 extreme value distribution, is one such distributions that deals with the extremely large values.

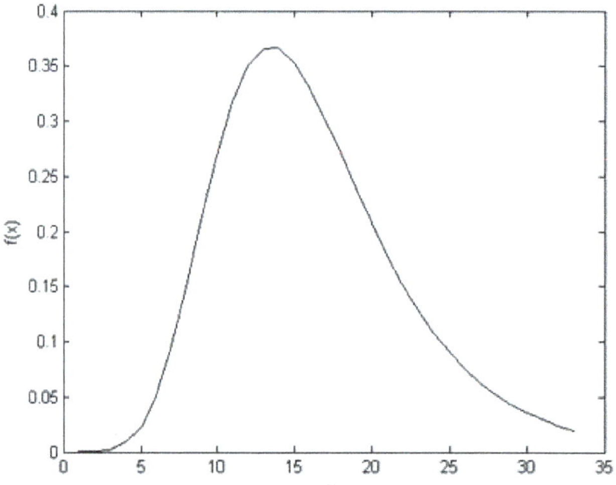

Gumbel distribution for the maximum extreme value with location parameter zero and the scale parameter 1

Gumbel Distribution

This distribution results from the underlying assumption that the elements of a random sample follow exponential distribution. Cumulative distribution function for the exponential distribution is shown below;

$$F(x) = 1 - e^{-\lambda x}$$

Then from the marginal distribution for the maximum value of the random sample which is equals to

$$Fx_{\max}(x) = [F(x)]^n$$

which results in,

$$Fx_{\max}(x) = [1 - e^{-\lambda x}]^n$$

With the introduction of the location and scale parameters b and a,

$$Fx_{\max}(x) = \left[1 - \frac{1}{n} e^{-\frac{x-b}{a}}\right]^n$$

As n tends to infinity it becomes,

$$Fx_{\max}(x) = \exp\left[e^{-\frac{x-b}{a}}\right]$$

The corresponding probability density function would be;

$$fx_{\max}(x) = \frac{1}{\alpha}\exp\left[-\frac{x-b}{\alpha} - e^{-\frac{x-b}{a}}\right] \text{ for } -\infty<x<\infty.$$

Gumbel distribution with location parameter 0 and the scale parameter is shown in the following figure.

Mean and the Variance of the Gumbel Distribution

The moment generating function for the Gumbel distribution is as follows;

$$Mx_{\max}(t) = \exp(bt)\Gamma(1-\alpha t) \; for \; t < 1/\alpha$$

The mean and the variance of the X_{max} are given below;

The mean, μ, equals to b + n_ea, where n_e is the Euler constant which is approximately equals to 0.5772.

The variance, σ^2 equals to $\frac{\pi^2\alpha}{6}$

If any random variable X follows Weibull distribution then the log X follows type 1 extreme value distribution known as the Gumbel distribution.

Properties of the Gumbel Distribution

1. The mode of this distribution is b

2. If X is Gumbel distributed with parameters (b, a) and l, and $m>0$ are scalar constants, $mX+l$ is Gumbel distributed with parameters

$$(m^*b+l, \frac{1}{\alpha * m})$$

3. If X_1, X_2 are two independent variables following Gumbel distribution with parameters (b_1, a) and (b_2, a), respectively, then the difference of these two variables, X^*, follow Logistic distribution.

$$F(X^*) = \frac{1}{1+e^{\frac{(b_2-b_1-X^*)}{\alpha}}}$$

4. If X_1, X_2 are two independent variables following Gumbel distribution with parameters (b_1, a) and (b_2, a), respectively, then the maximum of these two variables follows Gumbel distribution with the following parameters

$$\left(\alpha * In\left(e^{\frac{b_1}{\alpha}} + e^{\frac{b_2}{\alpha}}\right), 1/\alpha\right)$$

5. If X_1, X_2, ..., X_n are independent variables following Gumbel distribution with parameters (b_1, a) and (b_2, a), ... (b_n, a), respectively, then the maximum of these variables follows Gumbel distribution with the following parameters

$$\left(\alpha * In\sum_{i=1}^{n} e^{b_i/\alpha}, 1/\alpha\right)$$

These Properties are useful when dealing with the choice related data where the error/unobserved components are considered to follow the Gumbel distribution.

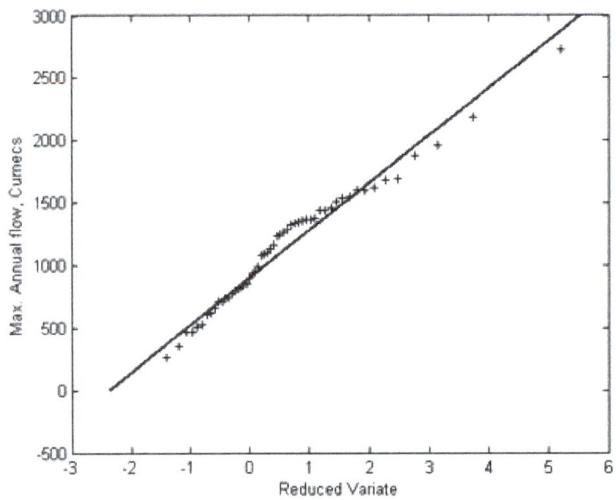

Gumbel probability plot for the annual maximum flows in the Tiber River, Italy

Fitting the Gumbel Distribution to the Annual Maximum Daily Runoff Data of a River

Daily runoff data is available for a river for about eighteen years. If only the maximum runoff data observed in a year is considered for modeling eighteen data points are available. Considering only the maximum runoff also obscures the other relatively larger daily runoffs occurred in any particular year. To avoid this, daily runoffs exceeding 500 Cumecs are considered in this example. This results in a total sample size of 120.

Table: Sample calculations involved in Gumbel plotting

S. No.	Run off (m^3/Sec)	Plotting position(p), m/n+0.3	Empirical reduced variate, -ln(-n(p))	Gumbel reduced variate, (x-b)/a
100	867.4	0.826	1.657	0.582
101	871.3	0.835	1.711	0.607
102	881.1	0.843	1.767	0.669
103	882.3	0.851	1.826	0.677
104	902.3	0.860	1.888	0.804
105	912.7	0.868	1.953	0.871
106	939.8	0.876	2.022	1.043
107	947.1	0.884	2.096	1.090
108	955.1	0.893	2.175	1.141
109	973.4	0.901	2.259	1.257
110	992.6	0.909	2.351	1.380
111	999.6	0.917	2.450	1.424
112	1006.2	0.926	2.560	1.466
113	1018.2	0.934	2.682	1.543
114	1034.2	0.942	2.820	1.645
115	1073.3	0.950	2.979	1.894
116	1128.6	0.959	3.165	2.246
117	1138	0.967	3.393	2.306
118	1307.2	0.975	3.685	3.383
119	1506.2	0.983	4.094	4.651
120	1528.6	0.992	4.792	4.794

Gumbel Probability Plotting

Two important steps in the Gumbel probability plotting are finding the plotting positions and the reduced variates. Plotting positions are distributed in between zero and one and each plotting position is equally likely to appear. To obtain the plotting positions the data is arranged in the ascending order and the ranks are given from the smallest to the largest variate. In finding the plotting position corresponding to any ordered statistic the rank value of that variate and the rank of the maxima is used. For example, the plotting position corresponding to the 100th ordered statistic is 100/(120+0.3). Sample calculations are shown in the above Table.

The parameters used in getting the theoretical reduced variates are: location parameter is 776, scale parameter is 157.

References

- Ugarte, MD; Militino, AF; Arnholt, AT (2016), Probability and Statistics with R (Second ed.), CRC Press, ISBN 978-1-4665-0439-4
- Johnson, N.L., Kotz, S., Kemp, A.W. (1993) Univariate Discrete distributions (2nd edition). Wiley. ISBN 0-471-54897-9, p153
- Clarke, R. D. (1946). "An application of the Poisson distribution" (PDF). Journal of the Institute of Actuaries. 72: 481
- Breslow, NE; Day, NE (1987). Statistical Methods in Cancer Research: Volume 2—The Design and Analysis of Cohort Studies. Paris: International Agency for Research on Cancer. ISBN 978-92-832-0182-3
- Johnson, N.L., Kotz, S., Kemp, A.W. (1993) Univariate Discrete distributions (2nd edition). Wiley. ISBN 0-471-54897-9, p157
- Joachim H. Ahrens; Ulrich Dieter (1982). "Computer Generation of Poisson Deviates". ACM Transactions on Mathematical Software. 8 (2): 163–179. doi:10.1145/355993.355997
- Ronald J. Evans; J. Boersma; N. M. Blachman; A. A. Jagers (1988). "The Entropy of a Poisson Distribution: Problem 87-6". SIAM Review. 30 (2): 314–317. doi:10.1137/1030059
- Riordan, John (1937). "Moment recurrence relations for binomial, Poisson and hypergeometric frequency distributions". Annals of Mathematical Statistics. 8: 103–111. doi:10.1214/aoms/1177732430
- Huiming, Zhang; Yunxiao Liu; Bo Li (2014). "Notes on discrete compound Poisson model with applications to risk theory". Insurance: Mathematics and Economics. 59: 325–336. doi:10.1016/j.insmatheco.2014.09.012
- Johnson, N.L., Kotz, S., Kemp, A.W. (1993) Univariate Discrete distributions (2nd edition). Wiley. ISBN 0-471-54897-9, p163
- Huiming, Zhang; Bo Li (2016). "Characterizations of discrete compound Poisson distributions". Communications in Statistics - Theory and Methods. 45: 6789–6802. doi:10.1080/03610926.2014.901375
- P.X., Gallagher (1976). "On the distribution of primes in short intervals". Mathematika. 23: 4–9. doi:10.1112/s0025579300016442
- Garwood, F. (1936). "Fiducial Limits for the Poisson Distribution". Biometrika. 28 (3/4): 437–442. doi:10.1093/biomet/28.3-4.437
- Michael Mitzenmacher & Eli Upfal. Probability and Computing: Randomized Algorithms and Probabilistic Analysis. Cambridge University Press. p. 97. ISBN 0521835402
- Edgeworth, F. Y. (1913). "On the use of the theory of probabilities in statistics relating to society". Journal of the Royal Statistical Society. 76: 165–193. JSTOR 10.2307/2340091
- Clevenson, M. L.; Zidek, J. V. (1975). "Simultaneous Estimation of the Means of Independent Poisson Laws". Journal of the American Statistical Association. 70 (351a): 698–705. doi:10.1080/01621459.1975.10482497
- Loukas, S.; Kemp, C. D. (1986). "The Index of Dispersion Test for the Bivariate Poisson Distribution". Biometrics. 42 (4): 941–948. doi:10.2307/2530708
- Philip J. Boland. "A Biographical Glimpse of William Sealy Gosset". The American Statistician, Vol. 38, No. 3. (Aug., 1984), pp. 179–183. Retrieved 2015-03-06
- Helske, Jouni (2015-06-25), KFAS: Exponential family state space models in R (PDF), Comprehensive R Archive Network, retrieved 2016-01-18

5

Sampling Distribution and Parameter Estimation

The population data is evaluated following probability distribution and verified through samples. The section explores several of these methods. In simple random sample, a subset is chosen from a larger set in a way such that all the elements have the same probability of getting selected. This chapter has been carefully written to provide an easy understanding of the varied facets of sample distribution.

Simple Random Sample

In statistics, a simple random sample is a subset of individuals (a sample) chosen from a larger set (a population). Each individual is chosen randomly and entirely by chance, such that each individual has the same probability of being chosen at any stage during the sampling process, and each subset of k individuals has the same probability of being chosen for the sample as any other subset of k individuals. This process and technique is known as simple random sampling, and should not be confused with systematic random sampling. A simple random sample is an unbiased surveying technique.

Simple random sampling is a basic type of sampling, since it can be a component of other more complex sampling methods. The principle of simple random sampling is that every object has the same probability of being chosen. For example, suppose N college students want to get a ticket for a basketball game, but there are only $X < N$ tickets for them, so they decide to have a fair way to see who gets to go. Then, everybody is given a number in the range from 0 to N-1, and random numbers are generated, either electronically or from a table of random numbers. Numbers outside the range from 0 to N-1 are ignored, as are any numbers previously selected. The first X numbers would identify the lucky ticket winners.

In small populations and often in large ones, such sampling is typically done "without replacement", i.e., one deliberately avoids choosing any member of the population more than once. Although simple random sampling can be conducted with replacement instead, this is less common and would normally be described more fully as simple random sampling with replacement. Sampling done without replacement is no longer independent, but still satisfies exchangeability, hence many results still hold. Further, for a small sample from a large population, sampling without replacement is approximately the same as sampling with replacement, since the odds of choosing the same individual twice is low.

An unbiased random selection of individuals is important so that if a large number of samples were drawn, the average sample would accurately represent the population. However, this does not guarantee that a particular sample is a perfect representation of the population. Simple random sampling merely allows one to draw externally valid conclusions about the entire population based on the sample.

Conceptually, simple random sampling is the simplest of the probability sampling techniques. It requires a complete sampling frame, which may not be available or feasible to construct for large populations. Even if a complete frame is available, more efficient approaches may be possible if other useful information is available about the units in the population.

Advantages are that it is free of classification error, and it requires minimum advance knowledge of the population other than the frame. Its simplicity also makes it relatively easy to interpret data collected in this manner. For these reasons, simple random sampling best suits situations where not much information is available about the population and data collection can be efficiently conducted on randomly distributed items, or where the cost of sampling is small enough to make efficiency less important than simplicity. If these conditions do not hold, stratified sampling or cluster sampling may be a better choice.

Algorithms

Several efficient algorithms for simple random sampling have been developed. A naive algorithm is the draw-by-draw algorithm where at each step we remove the item at that step from the set with equal probability and put the item in the sample. We continue until we have sample of desired size k. The drawback of this method is that it requires random access in the set.

The selection-rejection algorithm developed by Fan et al. in 1962 requires single pass over data; however, it is a sequential algorithm and requires knowledge of total count of items n, which is not available in streaming scenarios.

A very simple random sort algorithm was proved by Sunter in 1977 which simply assigns a random number drawn from uniform distribution (0, 1) as key to each item, sorts all items using the key and selects the smallest k items.

J. Vitter in 1985 proposed reservoir sampling algorithm which is often widely used. This algorithm does not require advance knowledge of n and uses constant space.

Random sampling can also be accelerated by sampling from the distribution of gaps between samples, and skipping over the gaps.

Distinction between a Systematic Random Sample and a Simple Random sample

Consider a school with 1000 students, and suppose that a researcher wants to select 100 of them for further study. All their names might be put in a bucket and then 100 names might be pulled out. Not only does each person have an equal chance of being selected, we can also easily calculate the probability P of a given person being chosen, since we know the sample size (n) and the population (N):

1. In the case that any given person can only be selected once (i.e., after selection a person is removed from the selection pool):

$$P = 1 - \frac{N-1}{N} \cdot \frac{N-2}{N-1} \cdot \cdots \cdot \frac{N-n}{N-(n-1)}$$

$$\overset{\text{Canceling:}}{=} 1 - \frac{N-n}{N}$$

$$= \frac{n}{N}$$
$$= \frac{100}{1000}$$
$$= 10\%$$

2. In the case that any selected person is returned to the selection pool (i.e., can be picked more than once):

$$P = 1 - \left(1 - \frac{1}{N}\right)^n = 1 - \left(\frac{999}{1000}\right)^{100} = 0.0952\ldots \approx 9.5\%$$

This means that every student in the school has in any case approximately a 1 in 10 chance of being selected using this method. Further, all combinations of 100 students have the same probability of selection.

If a systematic pattern is introduced into random sampling, it is referred to as "systematic (random) sampling". An example would be if the students in the school had numbers attached to their names ranging from 0001 to 1000, and we chose a random starting point, e.g. 0533, and then picked every 10th name thereafter to give us our sample of 100 (starting over with 0003 after reaching 0993). In this sense, this technique is similar to cluster sampling, since the choice of the first unit will determine the remainder. This is no longer simple random sampling, because some combinations of 100 students have a larger selection probability than others – for instance, {3, 13, 23, ..., 993} has a 1/10 chance of selection, while {1, 2, 3, ..., 100} cannot be selected under this method.

Sampling a Dichotomous Population

If the members of the population come in three kinds, say "blue" "red" and "black", the number of red elements in a sample of given size will vary by sample and hence is a random variable whose distribution can be studied. That distribution depends on the numbers of red and black elements in the full population. For a simple random sample *with* replacement, the distribution is a *binomial distribution*. For a simple random sample *without* replacement, one obtains a *hypergeometric distribution*.

- Multistage sampling
- Nonprobability sampling
- Opinion poll
- Quantitative marketing research

Random Samples and Statistics

In majority fields of Civil engineering it is necessary to deal with the sample data. For example, when modeling the travel behavior of the residents of a city it is difficult to approach each and every individual in the population and elicit the information on their travel pattern. A representative sample data is collected and based on that data the planners try to understand the travel behavior. Similarly, when deciding the characteristic strength of concrete, of a particular mix, data resulting from the tests on a sample is used. In case of finding the soil strength characteristics a sample number of bore-hole data is used. In all of these cases the sample is representing the population and while doing so the representative ability of the sample data must be ensured statistically.

Population data is assumed to follow certain probability distribution and this assumption is tested using the sample data. It is done by getting the estimates for the population distribution parameters. These estimates are obtained from the sample data.

Random Sample

X_1, X_2, X_3,......,X_n is a random sample of size n if X_i's are independent random variables and every observation X_i has the same probability distribution. If $f(x_1)$, $f(x_2)$, $f(x_3)$, ..., $f(x_n)$ are the probability distribution functions of the random variables of a random sample, the join probability distribution of the random sample is $f(x_1).f(x_2).f(x_3).....f(x_{n-1}).\ f(x_n)$.

Statistic

A statistic is the parameter related to the sample data. In other words, a statistic is any function of observations that is quantifiable and does not contain an unknown parameter. A statistic is useful in estimating the population parameters. Sample mean, sample standard deviation and the sample's variance are the statistics. Since the statistics are the functions of the random variables, they are also random variables.

Mean of a Random Sample

If X_1, X_2, X_3,......,X_n is a random sample then mean, $\bar{X} = \sum_{i=1\ to\ n} (X_i = x_i)_{/n}$.

Variance of the Random Sample

If X_1, X_2, X_3,......,X_n is a random sample then the variance, $S^2 = \sum_{i=1\ to\ n} (x_i = \bar{x})^2{}_{/n-1}$

Sampling Distribution

A random sample consists of several similar independent random variables and the joint probability distribution of these random variables is called the sampling distribution. The probability distribution of the mean of a random sample is a sampling distribution. Sampling distributions are useful in explaining the population data using sample parameters. In addition to this, they are useful in finding the sample size.

In finding the sampling distribution of a statistic there are several assumptions made, particularly on the population data from which the sample data is obtained. In many cases it is assumed that the population data is normally distributed. These statistics are known as parametric statistics.

An Example

In most of the traffic engineering problems it is necessary to have the knowledge of vehicular speed distributions. For example, if some upgradation work is carried on the road network the engineer would be interested in knowing the result of the upgradation. Before the upgradation process is initiated he must have had some objectives such as the increment in the overall speed would be 5 km/h. After the improvements in the network he may be interested in knowing whether the objectives are fulfilled or not. To know this it is necessary to have the speed data before and after the upgradation. In other words, sample speed data are necessary to test the level of improvement.

In this context, the term population refers to all the vehicles that are moving on the road network. It is assumed that the population speed data follow normal distribution. This assumption made on the population data is useful in knowing the distribution of the statistics or the sample parameters.

Nonparametric Statistics

Sometimes it is not possible to assume that the population data follow certain probability distribution, hence, not possible to arrive at the nature of the statistic. For example, in some experiments such as the observations on comfort, convenience etc. the outcomes cannot be quantified. In this case the statistic is called as a nonparametric statistic. Since no or few assumptions are made on the population data the applicability of nonparametric statistics is much wider.

One such statistic is the median of the sample data. This statistic is an alternative to the parametric statistic 'sample mean'.

Sampling Distributions

It is known that the speeds of a particular type of vehicle moving on more or less empty road, collected over several numbers of vehicles, follow normal distribution with parameters (μ, σ^2). But both the parameters are unknown and it is necessary to collect some sample speed data and based on the sample's information try to come to a decision about these parameters. In this scenario, sample data plays a major role in getting the estimates of the population parameters. Statistical or sampling distributions of the parameters of the sample data are useful in finding how close the sample parameters are to the population parameters. Using sampling distributions it is also possible to estimate the range or interval for population parameters with certain confidence.

Normal Distribution

Central limit theorem is useful in knowing the sampling distributions of different statistics. Using the definition of the random sample along with the central limit theorem it can be said that the mean of a relatively large sample ($\bar{X}$) follows normal distribution. The distribution of $\bar{X}$ is called as a sampling or statistical distribution. The mean of this sampling distribution is μ and the variance is σ^2/n. $\sigma/\sqrt{n}$ is known as the standard error of the sampling distribution. Even if the variance, σ^2, is unknown the sampling distribution follows normal distribution if the population data follow normal distribution or if the sample size is relatively large (> 30).

$X_1, X_2, \ldots X_n$ is the random sample of size n with mean $\bar{X}$ and the variance s^2 and the corresponding population parameters are μ and σ^2.

Since $\bar{X}$ is following normal distribution with mean μ and the variance σ^2/n, the following standardized random variable follows standard normal distribution.

$$Z = \frac{\overline{X} - \mu}{\sigma_{/\sqrt{n}}}$$

Sampling statistics such as the proportion of success or failures in case of a sequence of Bernoulli trials also follow normal distribution when n is greater than thirty. In this case the sampling distribution of the statistic $\hat{p}$ has the mean p and the standard error of $\sqrt{\frac{p(1-p)}{n}}$.

t – Distribution

If the variance (σ^2) is unknown and the sample is small in size (<30) then the sampling distribution of the sample mean follows t – distribution. If σ^2 is unknown, the sample›s variance can be used in place of it. Then the standard error of the sampling distribution of the sample mean is $S/\sqrt{n}$. If the sample size is greater than 30, even if the population variance is unknown, the sampling distribution of the sample mean can be considered to be normal distribution. Same is evident from Figures (a) and (b) in which it is shown that with the increasing sample size t-distribution converges to normal distribution.

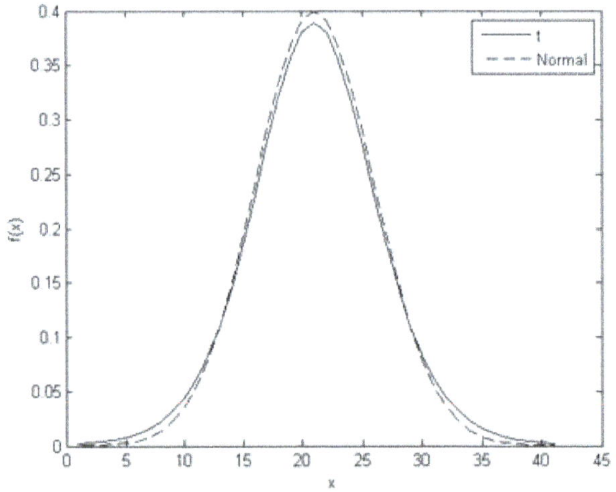

(a) Probability density functions of t-distribution (DOF=10) and normal distribution

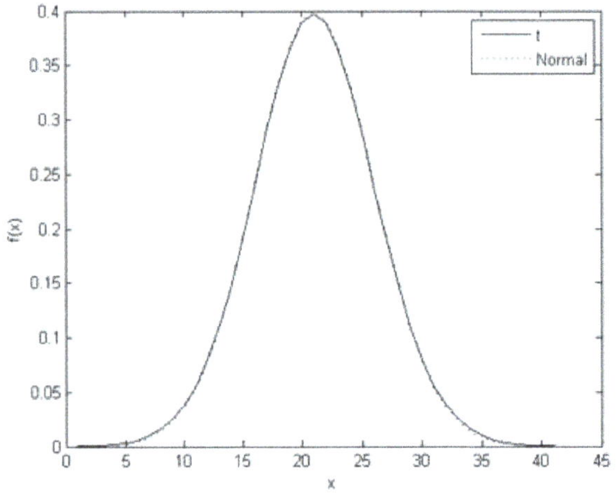

(b) Probability density functions of t-distribution (DOF=50) and normal distribution

Another example is if the sample size is less than thirty the proportion of success in a Bernoulli's experiment follows t-distribution.

If Z is a standard normal variable with mean zero and the variance one, and V is a chi-square random variable with k degrees of freedom and if both are independent, the resulting random variable T follows t-distribution with k degrees of freedom.

$$T = \frac{Z}{\sqrt{V/k}}$$

As discussed earlier, when the population data follows normal distribution and a random sample of size n is collected from the population, the sample's mean is μ and the variance is σ^2/n. When σ^2 is not known then it is replaced with S^2.

If the random variable $\frac{\bar{X}-\mu}{s/\sqrt{n}}$ is divided and multiplied with σ,

$$\frac{(\bar{X}-\mu)-\sigma}{s/\sqrt{n} * \sigma} \Rightarrow \frac{(\bar{X}-\mu)*\sigma}{\sigma/\sqrt{n} * s} \Rightarrow \frac{(\bar{X}-\mu)}{\sigma/\sqrt{n}} * \frac{1}{\sqrt{S^2/\sigma^2}}$$

It has been shown that $\frac{s^2}{\sigma^2}$ is equals to $\chi^2_{n-1}/(n-1)$.

As per the definition of the t-variable it can be said that $\frac{\bar{X}-\mu}{s/\sqrt{n}}$ follows t-distribution with n-1 degrees of freedom.

Chi-squared Distribution

In probability theory and statistics, the chi-squared distribution (also chi-square or χ^2-distribution) with k degrees of freedom is the distribution of a sum of the squares of k independent standard normal random variables. It is a special case of the gamma distribution and is one of the most widely used probability distributions in inferential statistics, e. g., in hypothesis testing or in construction of confidence intervals. When it is being distinguished from the more general noncentral chi-squared distribution, this distribution is sometimes called the central chi-squared distribution.

The chi-squared distribution is used in the common chi-squared tests for goodness of fit of an observed distribution to a theoretical one, the independence of two criteria of classification of qualitative data, and in confidence interval estimation for a population standard deviation of a normal distribution from a sample standard deviation. Many other statistical tests also use this distribution, like Friedman's analysis of variance by ranks.

Definition

If Z_1, ..., Z_k are independent, standard normal random variables, then the sum of their squares,

$$Q = \sum_{i=1}^{k} Z_i^2,$$

is distributed according to the chi-squared distribution with k degrees of freedom. This is usually denoted as

$$Q \sim \chi^2(k) \text{ or } Q \sim \chi_k^2.$$

The chi-squared distribution has one parameter: k — a positive integer that specifies the number of degrees of freedom (i. e. the number of Z_i's).

Introduction

The chi-squared distribution is used primarily in hypothesis testing. Unlike more widely known distributions such as the normal distribution and the exponential distribution, the chi-squared distribution is rarely used to model natural phenomena. It arises in the following hypothesis tests, among others.

- Chi-squared test of independence in contingency tables
- Chi-squared test of goodness of fit of observed data to hypothetical distributions
- Likelihood-ratio test for nested models
- Log-rank test in survival analysis
- Cochran–Mantel–Haenszel test for stratified contingency tables

It is also a component of the definition of the t-distribution and the F-distribution used in t-tests, analysis of variance, and regression analysis.

The primary reason that the chi-squared distribution is used extensively in hypothesis testing is its relationship to the normal distribution. Many hypothesis tests use a test statistic, such as the t statistic in a t-test. For these hypothesis tests, as the sample size, n, increases, the sampling distribution of the test statistic approaches the normal distribution (Central Limit Theorem). Because the test statistic (such as t) is asymptotically normally distributed, provided the sample size is sufficiently large, the distribution used for hypothesis testing may be approximated by a normal distribution. Testing hypotheses using a normal distribution is well understood and relatively easy. The simplest chi-squared distribution is the square of a standard normal distribution. So wherever a normal distribution could be used for a hypothesis test, a chi-squared distribution could be used.

Specifically, suppose that Z is a standard normal random variable, with mean = 0 and variance = 1. $Z \sim N(0,1)$. A sample drawn at random from Z is a sample from the distribution shown in the graph of the standard normal distribution. Define a new random variable Q. To generate a random sample from Q, take a sample from Z and square the value. The distribution of the squared values is given by the random variable $Q = Z^2$. The distribution of the random variable Q is an example of a chi-squared distribution: $Q \sim \chi_1^2$. The subscript 1 indicates that this particular chi-squared distribution is constructed from only 1 standard normal distribution. A chi-squared distribution constructed by squaring a single standard normal distribution is said to have 1 degree of freedom. Thus, as the sample size for a hypothesis test increases, the distribution of the test statistic approaches a normal distribution, and the distribution of the *square* of the test statistic approaches a chi-squared distribution. Just as extreme values of the normal distribution have low probability (and give small p-values), extreme values of the chi-squared distribution have low probability.

An additional reason that the chi-squared distribution is widely used is that it is a member of the class of likelihood ratio tests (LRT). LRT's have several desirable properties; in particular, LRT's commonly provide the highest power to reject the null hypothesis (Neyman–Pearson lemma). However, the normal and chi-squared approximations are only valid asymptotically. For this reason, it is preferable to use the t distribution rather than the normal approximation or the chi-squared approximation for small sample size. Similarly, in analyses of contingency tables, the chi-squared approximation will be poor for small sample size, and it is preferable to use Fisher's exact test. Ramsey and Ramsey show that the exact binomial test is always more powerful than the normal approximation.

Lancaster shows the connections among the binomial, normal, and chi-squared distributions, as follows. De Moivre and Laplace established that a binomial distribution could be approximated by a normal distribution. Specifically they showed the asymptotic normality of the random variable

$$\chi = \frac{m - Np}{\sqrt{(Npq)}}$$

where m is the observed number of successes in N trials, where the probability of success is p, and $q = 1 - p$.

Squaring both sides of the equation gives

$$\chi^2 = \frac{(m - Np)^2}{(Npq)}$$

Using $N = Np + N(1 - p)$, $N = m + (N - m)$, and $q = 1 - p$, this equation simplifies to

$$\chi^2 = \frac{(m - Np)^2}{(Np)} + \frac{(N - m - Nq)^2}{(Nq)}$$

The expression on the right is of the form that Pearson would generalize to the form:

$$\chi^2 = \sum_{i=1}^{n} \frac{(O_i - E_i)^2}{E_i}$$

where

χ^2 = Pearson's cumulative test statistic, which asymptotically approaches a χ^2 distribution.

O_i = the number of observations of type i.

$E_i = Np_i$ = the expected (theoretical) frequency of type i, asserted by the null hypothesis that the fraction of type i in the population is p_i

n = the number of cells in the table.

In the case of a binomial outcome (flipping a coin), the binomial distribution may be approximated by a normal distribution (for sufficiently large n). Because the square of a standard normal distribution is the chi-squared distribution with one degree of freedom, the probability of a result such as 1 heads in 10 trials can be approximated either by the normal or the chi-squared distribution. However, many problems involve more than the two possible outcomes of a binomial, and instead require 3 or more categories, which leads to the multinomial distribution. Just as de Moivre and Laplace sought for and found the normal approximation to the binomial, Pearson sought for and found a multivariate normal approximation to the multinomial distribution. Pearson showed that the chi-squared distribution, the sum of multiple normal distributions, was such an approximation to the multinomial distribution.

Probability Density Function

The probability density function (pdf) of the chi-square distribution is

$$f(x;k) = \begin{cases} \dfrac{x^{(k/2-1)}e^{-x/2}}{2^{k/2}\Gamma\left(\dfrac{k}{2}\right)}, & x > 0; \\ 0, & \text{otherwise.} \end{cases}$$

where $\Gamma(k/2)$ denotes the Gamma function, which has closed-form values for integer k.

For derivations of the pdf in the cases of one, two and k degrees of freedom.

Differential Equation

The pdf of the chi-squared distribution is a solution to the following differential equation:

$$2xf'(x) + f(x)(-k + x + 2) = 0$$

Therefore, $$\frac{f'(x)}{f(x)} = \frac{k-2-x}{2x} = \frac{\frac{k}{2}-1}{x} - \frac{1}{2}$$

Integrating both sides, $\log(f(x)) = (\frac{k}{2} - 1)\log(x) - \frac{x}{2} + \log(C)$

Therefore, $f(x) = Cx^{k/2-1}e^{-x/2} = C2^{k/2}(x/2)^{k/2-1}e^{-x/2}/2$

Total probability (i.e. area under the curve y = f(x) from -Infinity to +Infinity) needs to be 1.

Therefore, Total Probability $P = C2^{k/2}\int_0^{\infty}(x/2)^{k/2-1}e^{-x/2}d(x/2) = C2^{k/2}\Gamma(k/2) = 1$

Hence, PDF $\chi_k^2 = f(x) = Cx^{k/2-1}e^{-x/2} = \frac{x^{k/2-1}e^{-x/2}}{2^{k/2}\Gamma(k/2)}$ (and 0 when x < 0)

Cumulative Distribution Function

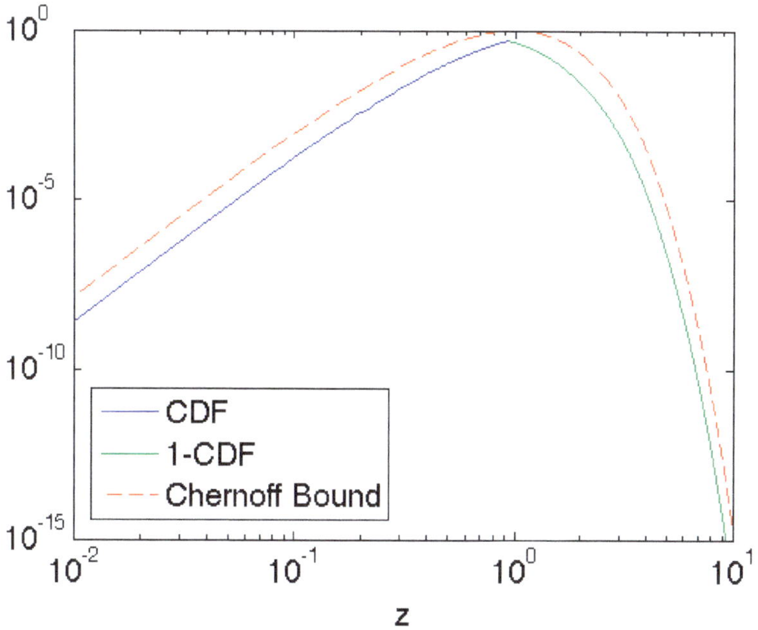

Chernoff bound for the CDF and tail (1-CDF) of a chi-squared random variable with ten degrees of freedom (k = 10)

Its cumulative distribution function is:

$$F(x;k) = \frac{(\frac{k}{2},\frac{x}{2})}{(-)} = P\left(\frac{k}{2},\frac{x}{2}\right),$$

where $\gamma(s,t)$ is the lower incomplete Gamma function and $P(s,t)$ is the regularized Gamma function.

In a special case of k = 2 this function has a simple form:

$$F(x;2) = 1 - e^{-x/2}$$

and the form is not much more complicated for other small even k.

Tables of the chi-squared cumulative distribution function are widely available and the function is included in many spreadsheets and all statistical packages.

Letting $z \equiv x/k$, Chernoff bounds on the lower and upper tails of the CDF may be obtained. For the cases when $0 < z < 1$ (which include all of the cases when this CDF is less than half):

$$F(zk;k) \le (ze^{1-z})^{k/2}.$$

The tail bound for the cases when $z > 1$, similarly, is

$$1 - F(zk;k) \le (ze^{1-z})^{k/2}.$$

For another approximation for the CDF modeled after the cube of a Gaussian.

Additivity

It follows from the definition of the chi-squared distribution that the sum of independent chi-squared variables is also chi-squared distributed. Specifically, if $\{X_i\}_{i=1}^n$ are independent chi-squared variables with $\{k_i\}_{i=1}^n$ degrees of freedom, respectively, then $Y = X_1 + \cdots + X_n$ is chi-squared distributed with $k_1 + \cdots + k_n$ degrees of freedom.

Sample Mean

The sample mean of n i.i.d. chi-squared variables of degree k is distributed according to a gamma distribution with shape α and scale θ parameters:

$$\overline{X} = \frac{1}{n}\sum_{i=1}^{n} X_i \sim \text{Gamma}\left(\alpha = nk/2, \theta = 2/n\right) \qquad \text{where } X_i \sim \chi^2(k)$$

Asymptotically, given that for a scale parameter α going to infinity, a Gamma distribution converges towards a normal distribution with expectation $\mu = \alpha \cdot \theta$ and variance $\sigma^2 = \alpha\theta^2$, the sample mean converges towards:

$$\overline{X} \xrightarrow{n\to\infty} N(\mu = k, \sigma^2 = 2k/n)$$

Note that we would have obtained the same result invoking instead the central limit theorem, noting that for each chi-squared variable of degree k the expectation is k, and its variance $2k$ (and hence the variance of the sample mean $\overline{X}$ being $\sigma^2 = 2k/n$).

Entropy

The differential entropy is given by

$$h = \int_{-\infty}^{\infty} f(x;k) \ln f(x;k)dx = \frac{k}{2} + \ln\left[2\Gamma\left(\frac{k}{2}\right)\right] + \left(1 - \frac{k}{2}\right)\psi\left[\frac{k}{2}\right],$$

where $\psi(x)$ is the Digamma function.

The chi-squared distribution is the maximum entropy probability distribution for a random variate X for which $\mathrm{E}(X)=k$ and $\mathrm{E}(\ln(X))=\psi(k/2)+\ln(2)$ are fixed. Since the chi-squared is in the family of gamma distributions, this can be derived by substituting appropriate values in the Expectation of the Log moment of Gamma. For derivation from more basic principles.

Noncentral Moments

The moments about zero of a chi-squared distribution with k degrees of freedom are given by

$$\mathrm{E}(X^m)=k(k+2)(k+4)\cdots(k+2m-2)=2^m\frac{\Gamma(m+\frac{k}{2})}{\Gamma(\frac{k}{2})}.$$

Cumulants

The cumulants are readily obtained by a (formal) power series expansion of the logarithm of the characteristic function:

$$\kappa_n=2^{n-1}(n-1)!k$$

Asymptotic Properties

By the central limit theorem, because the chi-squared distribution is the sum of k independent random variables with finite mean and variance, it converges to a normal distribution for large k. For many practical purposes, for $k > 50$ the distribution is sufficiently close to a normal distribution for the difference to be ignored. Specifically, if $X \sim \chi^2(k)$, then as k tends to infinity, the distribution of $(X-k)/\sqrt{2k}$ tends to a standard normal distribution. However, convergence is slow as the skewness is $\sqrt{8/k}$ and the excess kurtosis is $12/k$.

The sampling distribution of $\ln(\chi^2)$ converges to normality much faster than the sampling distribution of χ^2, as the logarithm removes much of the asymmetry. Other functions of the chi-squared distribution converge more rapidly to a normal distribution. Some examples are:

- If $X \sim \chi^2(k)$ then $\sqrt{2X}$ is approximately normally distributed with mean $\sqrt{2k-1}$ and unit variance (result credited to R. A. Fisher).
- If $X \sim \chi^2(k)$ then $\sqrt[3]{X/k}$ is approximately normally distributed with mean $1-2/(9k)$ and variance $2/(9k)$. This is known as the Wilson–Hilferty transformation.

Relation to other Distributions

- As $k\to\infty$, $(\chi_k^2-k)/\sqrt{2k}\xrightarrow{d}N(0,1)$ (normal distribution)
- $\chi_k^2 \sim \chi_k'^2(0)$ (noncentral chi-squared distribution with non-centrality parameter $\lambda=0$)
- If $Y \sim \mathrm{F}(\nu_1,\nu_2)$ then $X=\lim_{\nu_2\to\infty}\nu_1 Y$ has the chi-squared distribution $\chi_{\nu_1}^2$

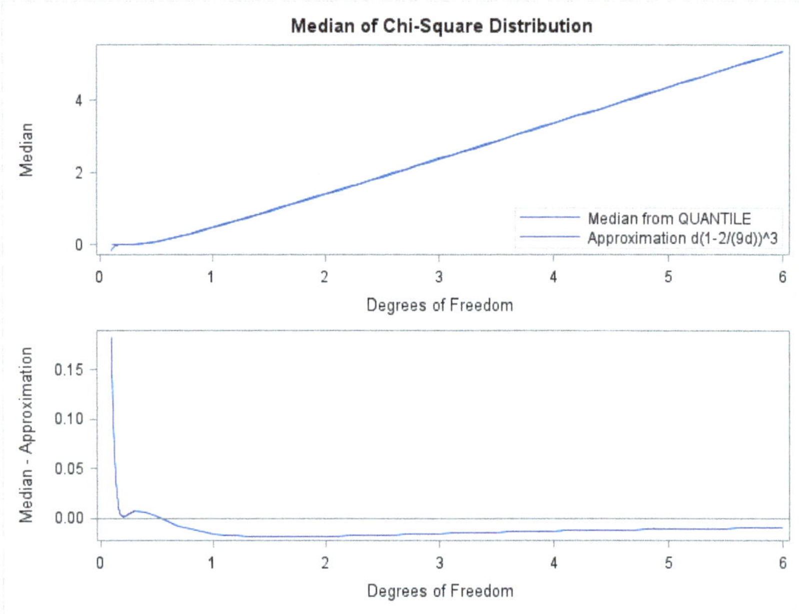

Approximate formula for median compared with numerical quantile (top). Difference between numerical quantile and approximate formula (bottom).

- ♦ As a special case, if $Y \sim \mathrm{F}(1,\nu_2)$ then $X = \lim_{\nu_2 \to \infty} Y$ has the chi-squared distribution χ_1^2
- $\| \mathbf{N}_{i=1,\dots,k}(0,1) \|^2 \sim \chi_k^2$ (The squared norm of k standard normally distributed variables is a chi-squared distribution with k degrees of freedom)
- If $X \sim \chi^2(\nu)$ and $c > 0$, then $cX \sim$ Ã$(k = \nu/2, \theta = 2c)$. (gamma distribution)
- If $X \sim \chi_k^2$ then $\sqrt{X} \sim \chi_k$ (chi distribution)
- If $X \sim \chi^2(2)$, then $X \sim \exp(1/2)$ is an exponential distribution.
- If $X \sim Rayleigh(1)$ (Rayleigh distribution) then $X^2 \sim \chi^2(2)$
- If $X \sim Maxwell(1)$ (Maxwell distribution) then $X^2 \sim \chi^2(3)$
- If $X \sim \chi^2(\nu)$ then $\frac{1}{X} \sim \text{Inv-}\chi^2(\nu)$ (Inverse-chi-squared distribution)
- The chi-squared distribution is a special case of type 3 Pearson distribution
- If $X \sim \chi^2(\nu_1)$ and $Y \sim \chi^2(\nu_2)$ are independent then $\frac{X}{X+Y} \sim Beta(\frac{\nu_1}{2}, \frac{\nu_2}{2})$ (beta distribution)
- If $X \sim \mathrm{U}(0,1)$ (uniform distribution) then $-2\log(X) \sim \chi^2(2)$
- $\chi^2(6)$ is a transformation of Laplace distribution
- If $X_i \sim Laplace(\mu,\beta)$ then $\sum_{i=1}^{n} \frac{2|X_i - \mu|}{\beta} \sim \chi^2(2n)$
- chi-squared distribution is a transformation of Pareto distribution

- Student's t-distribution is a transformation of chi-squared distribution
- Student's t-distribution can be obtained from chi-squared distribution and normal distribution
- Noncentral beta distribution can be obtained as a transformation of chi-squared distribution and Noncentral chi-squared distribution
- Noncentral t-distribution can be obtained from normal distribution and chi-squared distribution

A chi-squared variable with k degrees of freedom is defined as the sum of the squares of k independent standard normal random variables.

If Y is a k-dimensional Gaussian random vector with mean vector μ and rank k covariance matrix C, then $X = (Y-\mu)^T C^{-1}(Y - \mu)$ is chi-squared distributed with k degrees of freedom.

The sum of squares of statistically independent unit-variance Gaussian variables which do *not* have mean zero yields a generalization of the chi-squared distribution called the noncentral chi-squared distribution.

If Y is a vector of k i.i.d. standard normal random variables and A is a $k \times k$ symmetric, idempotent matrix with rank $k-n$ then the quadratic form $Y^T AY$ is chi-squared distributed with $k-n$ degrees of freedom.

The chi-squared distribution is also naturally related to other distributions arising from the Gaussian. In particular,

- Y is F-distributed, $Y \sim F(k_1,k_2)$ if $Y = \frac{X_1 / k_1}{X_2 / k_2}$ where $X_1 \sim \chi^2(k_1)$ and $X_2 \sim \chi^2(k_2)$ are statistically independent.
- If X is chi-squared distributed, then $\sqrt{X}$ is chi distributed.
- If $X_1 \sim \chi^2_{k1}$ and $X_2 \sim \chi^2_{k2}$ are statistically independent, then $X_1 + X_2 \sim \chi^2_{k1+k2}$. If X_1 and X_2 are not independent, then $X_1 + X_2$ is not chi-squared distributed.

Generalizations

The chi-squared distribution is obtained as the sum of the squares of k independent, zero-mean, unit-variance Gaussian random variables. Generalizations of this distribution can be obtained by summing the squares of other types of Gaussian random variables. Several such distributions are described below.

Linear Combination

If $X_1,\ldots,X_n$ are chi square random variables and $a_1,\ldots,a_n \in \mathbb{R}_{>0}$, then a closed expression for the distribution of $X = \sum_{i=1}^{n} a_i X_i$ is not known. It may be, however, calculated using the property of characteristic functions of the chi-squared random variable.

Chi-squared Distributions

Noncentral Chi-squared Distribution

The noncentral chi-squared distribution is obtained from the sum of the squares of independent Gaussian random variables having unit variance and *nonzero* means.

Generalized Chi-squared Distribution

The generalized chi-squared distribution is obtained from the quadratic form $z'Az$ where z is a zero-mean Gaussian vector having an arbitrary covariance matrix, and A is an arbitrary matrix.

Gamma, Exponential, and Related Distributions

The chi-squared distribution $X \sim \chi_k^2$ is a special case of the gamma distribution, in that $X \sim \Gamma\left(\frac{k}{2}, \frac{1}{2}\right)$ using the rate parameterization of the gamma distribution (or $X \sim \Gamma\left(\frac{k}{2}, 2\right)$ using the scale parameterization of the gamma distribution) where k is an integer.

Because the exponential distribution is also a special case of the Gamma distribution, we also have that if $X \sim \chi_2^2$, then $X \sim \exp\left(\frac{1}{2}\right)$ is an exponential distribution.

The Erlang distribution is also a special case of the Gamma distribution and thus we also have that if $X \sim \chi_k^2$ with even k, then X is Erlang distributed with shape parameter $k/2$ and scale parameter $1/2$.

Occurrence and Applications

The chi-squared distribution has numerous applications in inferential statistics, for instance in chi-squared tests and in estimating variances. It enters the problem of estimating the mean of a normally distributed population and the problem of estimating the slope of a regression line via its role in Student's t-distribution. It enters all analysis of variance problems via its role in the F-distribution, which is the distribution of the ratio of two independent chi-squared random variables, each divided by their respective degrees of freedom.

Following are some of the most common situations in which the chi-squared distribution arises from a Gaussian-distributed sample.

- if $X_1, \ldots, X_n$ are i.i.d. $N(\mu, \sigma^2)$ random variables, then $\sum_{i=1}^{n}(X_i - \bar{X})^2 \sim \sigma^2 \chi_{n-1}^2$ where $\bar{X} = \frac{1}{n}\sum_{i=1}^{n} X_i$.
- The box below shows some statistics based on $X_i \sim \text{Normal}(\mu_i, \sigma_i^2)$, $i = 1, \cdots, k$, independent random variables that have probability distributions related to the chi-squared distribution:

Name	Statistic
chi-squared distribution	$\sum_{i=1}^{k}\left(\frac{X_i-\mu_i}{\sigma_i}\right)^2$
noncentral chi-squared distribution	$\sum_{i=1}^{k}\left(\frac{X_i}{\sigma_i}\right)^2$
chi distribution	$\sqrt{\sum_{i=1}^{k}\left(\frac{X_i-\mu_i}{\sigma_i}\right)^2}$
noncentral chi distribution	$\sqrt{\sum_{i=1}^{k}\left(\frac{X_i}{\sigma_i}\right)^2}$

The chi-squared distribution is also often encountered in Magnetic Resonance Imaging.

Table of χ2 Values vs P-values

The *p*-value is the probability of observing a test statistic *at least* as extreme in a chi-squared distribution. Accordingly, since the cumulative distribution function (CDF) for the appropriate degrees of freedom *(df)* gives the probability of having obtained a value *less extreme* than this point, subtracting the CDF value from 1 gives the *p*-value. The table below gives a number of *p*-values matching to χ^2 for the first 10 degrees of freedom.

A low *p*-value indicates greater statistical significance, i. e., greater confidence that the observed deviation from the null hypothesis is significant. A *p*-value of 0.05 is often used as a cutoff between significant and not-significant results.

Degrees of freedom (df)	χ^2 value										
1	0.004	0.02	0.06	0.15	0.46	1.07	1.64	2.71	3.84	6.64	10.83
2	0.10	0.21	0.45	0.71	1.39	2.41	3.22	4.60	5.99	9.21	13.82
3	0.35	0.58	1.01	1.42	2.37	3.66	4.64	6.25	7.82	11.34	16.27
4	0.71	1.06	1.65	2.20	3.36	4.88	5.99	7.78	9.49	13.28	18.47
5	1.14	1.61	2.34	3.00	4.35	6.06	7.29	9.24	11.07	15.09	20.52
6	1.63	2.20	3.07	3.83	5.35	7.23	8.56	10.64	12.59	16.81	22.46
7	2.17	2.83	3.82	4.67	6.35	8.38	9.80	12.02	14.07	18.48	24.32
8	2.73	3.49	4.59	5.53	7.34	9.52	11.03	13.36	15.51	20.09	26.12
9	3.32	4.17	5.38	6.39	8.34	10.66	12.24	14.68	16.92	21.67	27.88
10	3.94	4.87	6.18	7.27	9.34	11.78	13.44	15.99	18.31	23.21	29.59
P value (Probability)	0.95	0.90	0.80	0.70	0.50	0.30	0.20	0.10	0.05	0.01	0.001

These values can be calculated evaluating the quantile function (also known as "inverse CDF" or "ICDF") of the chi-squared distribution; e. g., the χ^2 ICDF for $p = 1 - 0.95$ and df = 7 yields 14.067 ≈ 14.07 as in the table above. In Python, this can be calculated using SciPy via `scipy.stats.chi2.isf(0.05, 7)`.

History and Name

This distribution was first described by the German statistician Friedrich Robert Helmert in papers of 1875–6, where he computed the sampling distribution of the sample variance of a normal population. Thus in German this was traditionally known as the *Helmert'sche* ("Helmertian") or "Helmert distribution".

The distribution was independently rediscovered by the English mathematician Karl Pearson in the context of goodness of fit, for which he developed his Pearson's chi-squared test, published in 1900, with computed table of values published in (Elderton 1902), collected in (Pearson 1914, pp. xxxi–xxxiii, 26–28, Table XII). The name "chi-squared" ultimately derives from Pearson's shorthand for the exponent in a multivariate normal distribution with the Greek letter Chi, writing $-½\chi^2$ for what would appear in modern notation as $-½x^T\Sigma^{-1}x$ (Σ being the covariance matrix). The idea of a family of "chi-squared distributions", however, is not due to Pearson but arose as a further development due to Fisher in the 1920s.

The squared sum of the standard normal variables follows Chi-square distribution.

$$x^2 = z_1^2 + z_2^2 + z_3^2 + z_4^2 + \cdots.$$

Cumulative distribution function for the Chi-squared distribution is as follows;

$$F\left(\chi^2\right) = 1/2 \int_0^{\chi^2} \frac{\left(x/2\right)^{n/2^{-1}} e^{-x/2}}{r(\frac{n}{2})} dx$$

Where, Z_1, Z_2,..... are the standard normal variables.

X_1, X_2, ...X_n is the random sample of size n with mean $\bar{X}$ and variance s^2 and the corresponding population parameters are μ and σ^2.Then,

$$\frac{\sum_{i=1}^{n}\left(X_i-\mu\right)^2}{\sigma^2} \Rightarrow \frac{\sum_{i=1}^{n}\left(X_i-\bar{X}+\bar{X}-\mu\right)^2}{\sigma^2}$$

$$\Rightarrow \frac{\sum_{i=1}^{n}\left(X_i-\bar{X}\right)^2}{\sigma^2} + \frac{n\left(\bar{X}-\mu\right)}{\sigma^2}$$

$$\frac{\sum_{i=1}^{n}\left(X_i-\mu\right)^2}{\sigma^2} = \frac{\sum_{i=1}^{n}\left(X_i-\bar{X}\right)^2}{\sigma^2} + \left(\frac{\sqrt{n}\left(\bar{X}-\mu\right)}{\sigma}\right)^2$$

The left hand side of the above equation is the squared sum of standard normal variables and the resulting random variable follows Chi-Square distribution with n degrees of freedom. On the right hand side, the second part of the summation is the square of a standard normal variable and the resulting random variable follows Chi-square distribution with one degree of freedom. Resulting random variable of the summation of two Chi-squared random variables is a Chi-square random variable. Then, it can be said that the random variable in the first part of the summation is a Chi-square random variable with n-1 degrees of freedom. The figure below shows the probability density functions of the Chi-square random variables with various degrees of freedom.

The variance of the random sample is given below and from the variance it is shown that the random variable $(n-1)\left(\frac{s^2}{\sigma^2}\right)$ follows Chi-square distribution with n-1 degrees of freedom.

$$s^2 = \frac{\sum_{i=1}^{n}\left(X_i - \bar{X}\right)^2}{n-1}$$

$$(n-1)\frac{s^2}{\sigma^2} = \frac{\sum_{i=1}^{n}\left(X_i - \bar{X}\right)^2}{\sigma^2}$$

$$(n-1)\frac{s^2}{\sigma^2} = Z_1^2 + Z_2^2 + Z_3^2 + Z_4^2 + \cdots + Z_n^2.$$

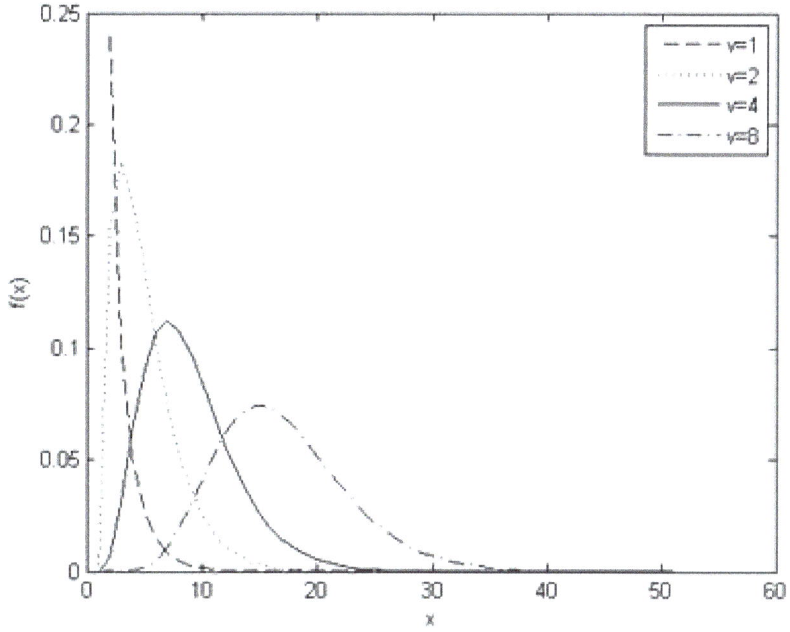

Chi-square pdf with various degrees of freedom

The F-Distribution

If χ_m^2 and χ_n^2 are two independent Chi-square random variables with m and n degrees of freedom, respectively, then the following random variable follows F-distribution with m and ndegrees of freedom.

$$F_{m,n} = \frac{\chi_m^2 / m}{\chi_n^2 / n}$$

From the previous discussion on Chi-square distribution it is clear that $(n-1)\frac{s^2}{\sigma^2}$ follows Chi-square distribution. Then, if two random samples of same size, n, are drawn from the two populations and the sample and population variances are s_1^2, σ_1^2 and s_2^2, σ_2^2, respectively, the following random variable follows F-Distribution with n-1, and n-1 degrees of freedom;

$$F = \frac{s_1^2 / \sigma_1^2}{s_2^2 / \sigma_2^2}$$

This distribution is particularly useful in comparing the variances of two populations using the random samples obtained from them. The figure below shows the probability density functions of a random variable following F-distribution with various degrees of freedom.

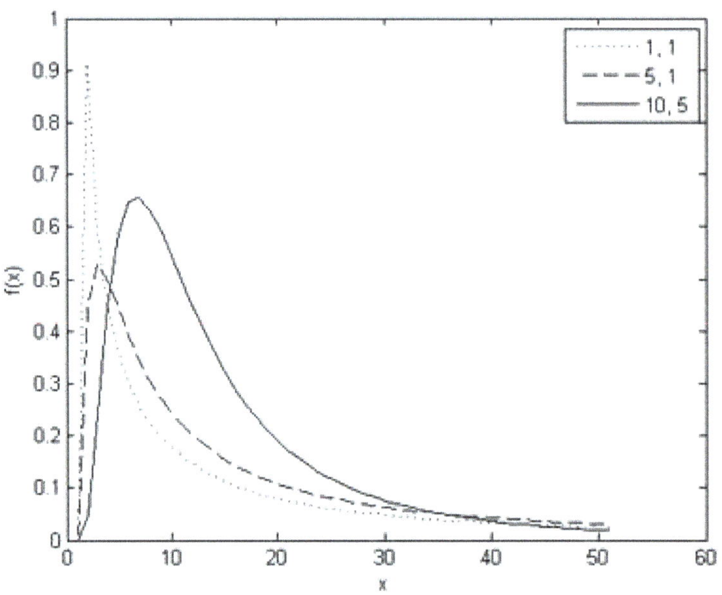

F-distribution pdf for various degrees of freedom

Estimation Theory

Estimation theory is a branch of statistics that deals with estimating the values of parameters based on measured empirical data that has a random component. The parameters describe an underlying physical setting in such a way that their value affects the distribution of the measured data. An estimator attempts to approximate the unknown parameters using the measurements.

For example, it is desired to estimate the proportion of a population of voters who will vote for a particular candidate. That proportion is the parameter sought; the estimate is based on a small random sample of voters.

Or, for example, in radar Our aim is to find the range of objects (airplanes, boats, etc.) by analyzing the two-way transit timing of received echoes of transmitted pulses. Since the reflected pulses are unavoidably embedded in electrical noise, their measured values are randomly distributed, so that the transit time must be estimated.

In estimation theory, two approaches are generally considered.

- The probabilistic approach assumes that the measured data is random with probability distribution dependent on the parameters of interest
- The set-membership approach assumes that the measured data vector belongs to a set which depends on the parameter vector.

For example, in electrical communication theory, the measurements which contain information regarding the parameters of interest are often associated with a noisy signal. Without randomness, or noise, the problem would be deterministic and estimation would not be needed.

Basics

To build a model, several statistical "ingredients" need to be known. These are needed to ensure the estimator has some mathematical tractability.

The first is a set of statistical samples taken from a random vector (RV) of size N. Put into a vector,

$$\mathbf{x} = \begin{bmatrix} x[0] \\ x[1] \\ \vdots \\ x[N-1] \end{bmatrix}.$$

Secondly, there are the corresponding M parameters

$$\theta = \begin{bmatrix} \theta_1 \\ \theta_2 \\ \vdots \\ \theta_M \end{bmatrix},$$

which need to be established with their continuous probability density function (pdf) or its discrete counterpart, the probability mass function (pmf)

$$\mathrm{p}(\mathrm{x} \mid \theta).$$

It is also possible for the parameters themselves to have a probability distribution (e.g., Bayesian statistics). It is then necessary to define the Bayesian probability

$$\pi(\theta).$$

After the model is formed, the goal is to estimate the parameters, commonly denoted $\hat{\theta}$, where the "hat" indicates the estimate.

One common estimator is the minimum mean squared error estimator, which utilizes the error between the estimated parameters and the actual value of the parameters

$$e = \hat{\theta} - \theta$$

as the basis for optimality. This error term is then squared and minimized for the MMSE estimator.

Estimators

Commonly used estimators and estimation methods, and topics related to them:

- Maximum likelihood estimators
- Bayes estimators
- Method of moments estimators
- Cramér–Rao bound
- Minimum mean squared error (MMSE), also known as Bayes least squared error (BLSE)
- Maximum a posteriori (MAP)
- Minimum variance unbiased estimator (MVUE)
- nonlinear system identification
- Best linear unbiased estimator (BLUE)
- Unbiased estimators
- Particle filter
- Markov chain Monte Carlo (MCMC)
- Kalman filter, and its various derivatives
- Wiener filter

Examples

Unknown Constant in Additive White Gaussian Noise

Consider a received discrete signal, $x[n]$, of N independent samples that consists of an unknown constant A with additive white Gaussian noise (AWGN) $w[n]$ with known variance σ^2 (*i.e.*, $\mathcal{N}(0,\sigma^2)$). Since the variance is known then the only unknown parameter is A.

The model for the signal is then

$$x[n] = A + w[n] \quad n = 0,1,\ldots,N-1$$

Two possible (of many) estimators for the parameter A are:

- $\hat{A}_1 = x[0]$
- $\hat{A}_2 = \frac{1}{N}\sum_{n=0}^{N-1} x[n]$ which is the sample mean

Both of these estimators have a mean of A, which can be shown through taking the expected value of each estimator

$$\mathrm{E}\left[\hat{A}_1\right] = \mathrm{E}\left[x[0]\right] = A$$

and

$$\mathrm{E}\left[\hat{A}_2\right] = \mathrm{E}\left[\frac{1}{N}\sum_{n=0}^{N-1} x[n]\right] = \frac{1}{N}\left[\sum_{n=0}^{N-1} \mathrm{E}\left[x[n]\right]\right] = \frac{1}{N}[NA] = A$$

At this point, these two estimators would appear to perform the same. However, the difference between them becomes apparent when comparing the variances.

$$\mathrm{var}\left(\hat{A}_1\right) = \mathrm{var}\left(x[0]\right) = \sigma^2$$

and

$$\mathrm{var}\left(\hat{A}_2\right) = \mathrm{var}\left(\frac{1}{N}\sum_{n=0}^{N-1} x[n]\right) \overset{\text{independence}}{=} \frac{1}{N^2}\left[\sum_{n=0}^{N-1} \mathrm{var}(x[n])\right] = \frac{1}{N^2}\left[N\sigma^2\right] = \frac{\sigma^2}{N}$$

It would seem that the sample mean is a better estimator since its variance is lower for every $N > 1$.

Maximum Likelihood

Continuing the example using the maximum likelihood estimator, the probability density function (pdf) of the noise for one sample $w[n]$ is

$$p(w[n]) = \frac{1}{\sigma\sqrt{2\pi}}\exp\left(-\frac{1}{2\sigma^2}w[n]^2\right)$$

and the probability of $x[n]$ becomes ($x[n]$ can be thought of a $\mathcal{N}(A,\sigma^2)$)

$$p(x[n];A) = \frac{1}{\sigma\sqrt{2\pi}}\exp\left(-\frac{1}{2\sigma^2}(x[n]-A)^2\right)$$

By independence, the probability of $\mathbf{x}$ becomes

$$p(\mathbf{x};A)=\prod_{n=0}^{N-1}p(x[n];A)=\frac{1}{\left(\sigma\sqrt{2\pi}\right)^N}\exp\left(-\frac{1}{2\sigma^2}\sum_{n=0}^{N-1}(x[n]-A)^2\right)$$

Taking the natural logarithm of the pdf

$$\ln p(\mathbf{x};A)=-N\ln\left(\sigma\sqrt{2\pi}\right)-\frac{1}{2\sigma^2}\sum_{n=0}^{N-1}(x[n]-A)^2$$

and the maximum likelihood estimator is

$$\hat{A}=\arg\max \ln p(\mathbf{x};A)$$

Taking the first derivative of the log-likelihood function

$$\frac{\partial}{\partial A}\ln p(\mathbf{x};A)=\frac{1}{\sigma^2}\left[\sum_{n=0}^{N-1}(x[n]-A)\right]=\frac{1}{\sigma^2}\left[\sum_{n=0}^{N-1}x[n]-NA\right]$$

and setting it to zero

$$0=\frac{1}{\sigma^2}\left[\sum_{n=0}^{N-1}x[n]-NA\right]=\sum_{n=0}^{N-1}x[n]-NA$$

This results in the maximum likelihood estimator

$$\hat{A}=\frac{1}{N}\sum_{n=0}^{N-1}x[n]$$

which is simply the sample mean. From this example, it was found that the sample mean is the maximum likelihood estimator for N samples of a fixed, unknown parameter corrupted by AWGN.

Cramér–Rao Lower Bound

To find the Cramér–Rao lower bound (CRLB) of the sample mean estimator, it is first necessary to find the Fisher information number

$$\mathcal{I}(A)=\mathrm{E}\left(\left[\frac{\partial}{\partial A}\ln p(\mathbf{x};A)\right]^2\right)=-\mathrm{E}\left[\frac{\partial^2}{\partial A^2}\ln p(\mathbf{x};A)\right]$$

and copying from above

$$\frac{\partial}{\partial A}\ln p(\mathbf{x};A)=\frac{1}{\sigma^2}\left[\sum_{n=0}^{N-1}x[n]-NA\right]$$

Taking the second derivative

$$\frac{\partial^2}{\partial A^2}\ln p(\mathbf{x};A) = \frac{1}{\sigma^2}(-N) = \frac{-N}{\sigma^2}$$

and finding the negative expected value is trivial since it is now a deterministic constant

$$-\mathrm{E}\left[\frac{\partial^2}{\partial A^2}\ln p(\mathbf{x};A)\right] = \frac{N}{\sigma^2}$$

Finally, putting the Fisher information into

$$\mathrm{var}\left(\hat{A}\right) \geq \frac{1}{\mathcal{I}}$$

results in

$$\mathrm{var}\left(\hat{A}\right) \geq \frac{\sigma^2}{N}$$

Comparing this to the variance of the sample mean (determined previously) shows that the sample mean is *equal to* the Cramér–Rao lower bound for all values of N and A. In other words, the sample mean is the (necessarily unique) efficient estimator, and thus also the minimum variance unbiased estimator (MVUE), in addition to being the maximum likelihood estimator.

Maximum of a Uniform Distribution

One of the simplest non-trivial examples of estimation is the estimation of the maximum of a uniform distribution. It is used as a hands-on classroom exercise and to illustrate basic principles of estimation theory. Further, in the case of estimation based on a single sample, it demonstrates philosophical issues and possible misunderstandings in the use of maximum likelihood estimators and likelihood functions.

Given a discrete uniform distribution $1, 2, \ldots, N$ with unknown maximum, the UMVU estimator for the maximum is given by

$$\frac{k+1}{k}m - 1 = m + \frac{m}{k} - 1$$

where m is the sample maximum and k is the sample size, sampling without replacement. This problem is commonly known as the German tank problem, due to application of maximum estimation to estimates of German tank production during World War II.

The formula may be understood intuitively as;

> "The sample maximum plus the average gap between observations in the sample",

the gap being added to compensate for the negative bias of the sample maximum as an estimator for the population maximum.

This has a variance of

$$\frac{1}{k}\frac{(N-k)(N+1)}{(k+2)} \approx \frac{N^2}{k^2} \text{ for small samples } k \ll N$$

so a standard deviation of approximately N/k, the (population) average size of a gap between samples; compare $\frac{m}{k}$ above. This can be seen as a very simple case of maximum spacing estimation.

The sample maximum is the maximum likelihood estimator for the population maximum, but, as discussed above, it is biased.

Applications

Numerous fields require the use of estimation theory. Some of these fields include (but are by no means limited to):

- Interpretation of scientific experiments
- Signal processing
- Clinical trials
- Opinion polls
- Quality control
- Telecommunications
- Project management
- Software engineering
- Control theory (in particular Adaptive control)
- Network intrusion detection system
- Orbit determination

Measured data are likely to be subject to noise or uncertainty and it is through statistical probability that optimal solutions are sought to extract as much information from the data as possible.

Parameter Estimation

Parameter estimation is crucial in statistical decision making. It is common practice in civil engineering to collect sample data to estimate the parameters of population distribution. The parameters estimated from the sample data and their sampling distributions are useful in estimating the population parameters. Two types of estimates are possible from the sample data, namely, point estimates, interval estimates.

Properties of the Estimators

An *estimator* is the rule or the method of estimation and an *estimate* is the value the estimator yields in a particular case. There may be several estimators but an appropriate estimator must have the following properties.

1. Unbiasedness
2. Consistency
3. Minimum Variability
4. Efficiency
5. Sufficiency

When the expectation of the unbiased estimator yields the corresponding population parameter then the estimator is known as unbiased estimator. The difference between the expectation of the unbiased estimator and the corresponding population parameter is known as *bias*.

For example $\bar{X}$ is the sample mean and if $\ddot{\bar{X}}$ is to be the unbiased estimator for the population mean μ $\left(\hat{\bar{X}}\right)$ must be equal to μ. In this case the estimate for the population mean is $\bar{X}$.

The estimator must also have minimum variance when it is obtained through repeated sampling. A *sufficient estimator* provides as much information as possible on the sample data and better than any other estimator.

Point Estimators

X_1, X_2, X_3, ... X_n is a random sample of size n collected from the population with parameters μ and σ^2. The sample has mean $\bar{X}$ and variance S^2. In this case the unbiased point estimators for the population parameters are $\ddot{\bar{X}}$ and $\hat{s}^2$. Point estimates may lead to errors when the variance associated to the estimators is high. To avoid this, confidence intervals can be provided for the population parameters with the help of the sampling distributions.

Parameter Estimation Methods

Maximum Likelihood Estimation Technique

In most of the cases MLE technique is used in estimating the parameters. If X is a random variable with the probability density function *f*(*x*) and α is the parameter (similar to the single parameter exponential distribution where λ is the only parameter), its value can be estimated using a sample data set by formulating the likelihood function. If the sample size is n, in this case the likelihood function is as follows;

$$L(\alpha) = f(x_1,\ \alpha).f(x_2,\ \alpha).f(x_3,\ \alpha)\ \ldots\ldots f(x_n,\ \alpha)$$

The maximum likelihood parameter of the α is the value α that maximizes the likelihood function L(α).

Using MLE technique, If the population data is assumed to follow the Bernoulli distribution, the parameter p of the distribution can be found using a random sample of size n collected from the population data, as shown below.

The probability mass function of the Bernoulli variable is $p(x) = p^x (1-p)^{(1-x)}$

The likelihood function resulting from the sample of size n is;

$$L(p) = \prod_{i=1}^{n} p^{xi} (1-P)^{(1-xi)}$$

Log likelihood function can be written as follows;

$$In\ L(p) = \sum_{i=1}^{n} (In\ p^{xi} + In(1+P)^{(1-xi)})$$

$$In\ L(p) = \sum_{i=1}^{n} (x_i In\ p + (1-x_i) In(1-P))$$

$$\frac{d}{dp} In\ L(p) = \frac{\sum_{i=1}^{n} x_i}{p} - \frac{\left(n - \sum_{i=1}^{n} x_i\right)}{1-p}$$

Corresponding to the maximum log likelihood the above differential is equal to zero, then;

$$\hat{p} = \frac{1}{n} \sum_{i=1}^{n} x_i$$

The parameters obtained from the MLE technique may not be unbiased always. One way to avoid the bias is to increase the sample size used in the parameter estimation. Maximum likelihood estimators are always consistent.

Method of Moments

This method of parameter estimation is based on the origin moments of the random variable. Moments about origin are to be calculated from the sample data and equate them to the corresponding theoretical moments of the assumed distribution. For example, if some population data is assumed to follow normal distribution, the first two moments about the origin are μ and $\mu^2 + \sigma^2$. Equating these moments to the origin moments calculated from the sample of size nresults in the following two equations.

$$\mu = \frac{1}{n} \sum_{i=1}^{n} X_i$$

$$\sigma^2 + \mu^2 = \frac{1}{n} \sum_{i=1}^{n} X_i^2$$

Solving the above two equations results in the following estimators for μ and σ^2,

$$\hat{\mu} = \frac{1}{n}\sum_{i=1}^{n} X_i = \bar{X}$$

$$\hat{\sigma}^2 = \frac{1}{n}\sum_{i=1}^{n}\left(X_i - \bar{X}\right)^2$$

Often the estimators from the method of moments are reasonably good. The variance of the estimators resulting from the method of moments are larger compared to the estimators resulting from the other methods.

Confidence Interval Estimation

As discussed earlier it is also possible to estimate the confidence intervals for the population parameters in terms of the sample parameters. This requires knowledge of the sampling distributions of the sample parameters. It deals with the determination of confidence intervals for the population mean, population variance when the population data is approximately normally distributed. Another section deals with the estimation of confidence interval for the proportion of favorable outcomes (related to an experiment) in the population data.

Confidence Interval for the Population Mean (μ), when the Population Variance is Known

Confidence interval for any particular population parameter denotes the range in which the population parameter falls under repetitive sampling, at a particular confidence level. The confidence level denotes the probability corresponding to the area under the corresponding probability density function, encompassed by the upper and lower limits of the confidence interval. If the area under the *pdf* bound by the confidence interval is 0.95 the corresponding confidence level is 95%. The remaining 5 % is known as the *significance*. The following figure shows the graphical representation of the two-sided confidence interval corresponding to 95% confidence.

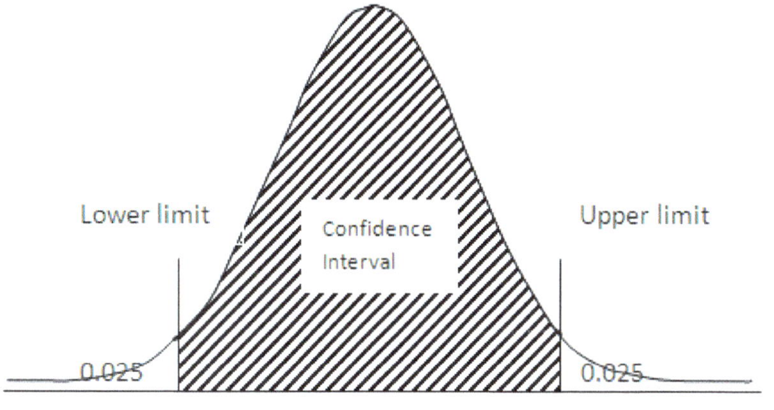

Area under the normal curve for 95% confidence level with lower and upper limits

This is estimated from the random sample, $X_1, X_2, X_3, ... X_n$, of size n collected from the population with parameters μ and σ^2. In this case the population variance σ^2 is known. The sample has mean $\bar{X}$ and variance S^2. It is known that the sample mean $\bar{X}$ is normally distributed with mean μ and variance σ^2/n. Then a transformed variable Z follows standard normal distribution.

$$Z = \frac{\bar{X} - \mu}{\sigma / \sqrt{n}}$$

Given a significance level of α, the confidence interval for μ can be obtained as follows; from the previous discussion it is known that area under the pdf bound between the upper and lower limits is 1- α. The same can be expressed as follows;

$$P\left\{ -z\alpha/_2 \leq \frac{\bar{X} - \mu}{\sigma / \sqrt{n}} \leq z\alpha/_2 \right\} = 1 - \alpha$$

$$P\left\{ -z\alpha/_2 \, \sigma/\sqrt{n} \leq \bar{X} - \mu \leq z\alpha/_2 \, \sigma/\sqrt{n} \right\} = 1 - \alpha$$

$$P\left\{ \bar{X} - z\alpha/_2 \, \sigma/\sqrt{n} \leq \mu \leq \bar{X} \; z\alpha/_2 \, \sigma/\sqrt{n} \right\} = 1 - \alpha$$

$$P\left\{ \bar{X} - z\alpha/_2 \, Se(\bar{X}) \leq \mu \leq \bar{X} \; z\alpha/_2 \, Se(\bar{X}) \right\} = 1 - \alpha$$

Then the 100(1-α) confidence interval for the population mean is

$$\bar{X} - z\alpha/_2 \, Se(\bar{X}) \leq \mu \leq \bar{X} \; z\alpha/_2 \, \sigma/\sqrt{n}$$

Now the upper and lower limits for μ are obtained in terms of samples statistic and its standard error.

Similarly upper bound or lower bound (the other limit being infinity) can be found. When the lower limit is to be obtained the critical region lies in the lower tail of the probability distribution. This is also known as the upper confidence interval. The following figure is the graphical representation of the upper confidence interval.

$$\bar{X} - z_\alpha \, \sigma/\sqrt{n} \leq \mu$$

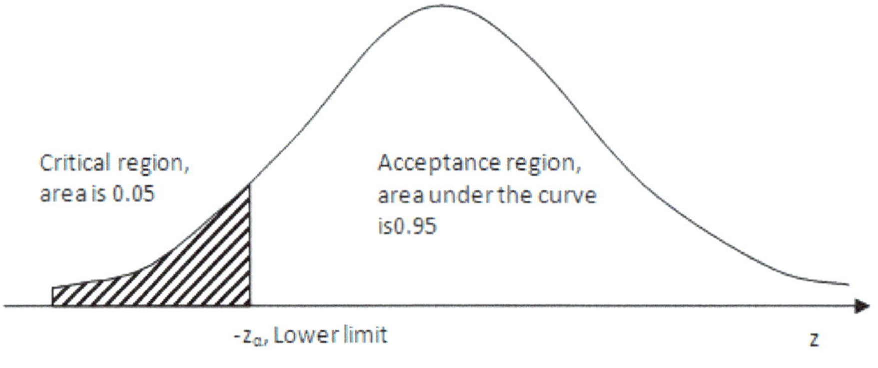

Area under the normal curve for 95% confidence level with lower limit

When the upper limit is to be obtained the critical region lies in the upper tail of the probability distribution. The lower limit is -∞. This is known as the lower confidence interval.

The figure below is the graphical representation of the lower confidence interval.

$$\mu \leq \bar{X} + z_\alpha \sigma/\sqrt{n}$$

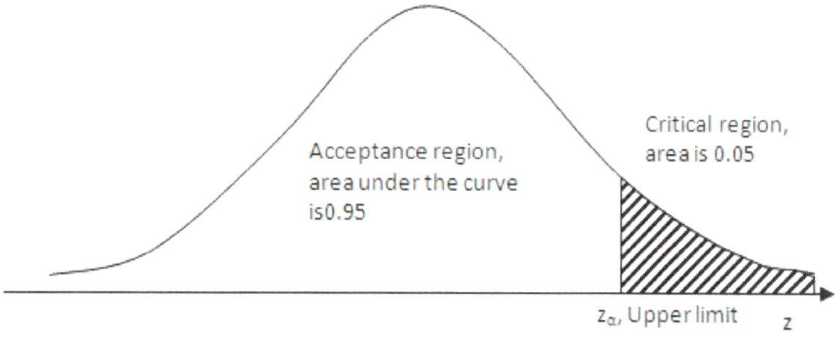

Area under the normal curve for 95% confidence level with upper limit

Confidence Interval for the Population Mean (μ) when the Population Variance is Unknown

In this case the population variance σ^2 is unknown. The sample has mean $\bar{X}$ and variance S^2. It is known that the sample mean $\bar{X}$ is normally distributed with mean μ and variance σ^2/n. Since the variance is unknown the sample variance is used in place of the population variance. Then a transformed variable or the test statistic follows t- distribution.

$$t = \frac{\bar{X} - \mu}{s/\sqrt{n}}$$

Given a significance level of α, the confidence interval for μ can be obtained. It is known that the area under the pdf bound between the upper and lower limits is 1- α. The same can be expressed as follows;

$$P\left\{-t_{\alpha/2} \leq \frac{\bar{X} - \mu}{S/\sqrt{n}} \leq t_{\alpha/2}\right\} = 1 - \alpha$$

$$P\left\{-t_{\alpha/2} S/\sqrt{n} \leq \bar{X} - \mu \leq t_{\alpha/2} S/\sqrt{n}\right\} = 1 - \alpha$$

$$P\left\{\bar{X} - t_{\alpha/2} S/\sqrt{n} \leq \mu \leq \bar{X} + t_{\alpha/2} S/\sqrt{n}\right\} = 1 - \alpha$$

$$P\left\{\bar{X} - t_{\alpha/2} Se\left(\bar{X}\right) \leq \mu \leq \bar{X} + t_{\alpha/2} Se\left(\bar{X}\right)\right\} = 1 - \alpha$$

Then the 100(1-α) confidence interval for the population mean is

$$\bar{X} - t_{\alpha/2} S/\sqrt{n} \le \mu \le \bar{X} + t_{\alpha/2} S/\sqrt{n})$$

When the sample size is greater than 30, even if the variance is unknown the test statistic may be assumed to follow normal distribution. In this case the 100(1-α) confidence interval would be as follows;

$$\bar{X} - Z_{\alpha/2} S/\sqrt{n} \le \mu \le \bar{X} + Z_{\alpha/2} S/\sqrt{n}$$

Problem

Establish a 98 % confidence interval for the mean pile strength. Assume that the variance is known (take the sample variance as the population variance) and unknown.

Answer:

Confidence interval on mean when the variance is known:
Sample size = 9
Sample mean = 85 tons
Population standard deviation is the sample standard deviation = 6.76 tons
Significance level = 2%
Two – sided confidence interval:
Since the population variance is known the sampling statistic follows normal distribution. In this case the two sided confidence interval is,

$$\bar{X} - Z_{\alpha/2} S/\sqrt{n} \le \mu \le \bar{X} + Z_{\alpha/2} S/\sqrt{n}$$

For $z_{0.01} = 2.33$ the confidence interval is

$7.85 \le \mu \le 91.5$

Confidence interval on mean when the variance is unknown:
Sample size = 9
Sample mean = 85 tons
Sample standard deviation = 6.76 tons
Significance level = 2%
Two – sided confidence interval:
Since the population variance is unknown the sampling statistic follows t – distribution. In this case the two sided confidence interval is,

$$\bar{X} - t_{\alpha/2} S/\sqrt{n} \le \mu \le \bar{X} + t_{\alpha/2} S/\sqrt{n}$$

For $t_{0.01, 8} = 2.8965$ the confidence interval is

$7.85 \le \mu \le 91.5$

Confidence Interval on the Difference of two Population Means (Known Population Variances)

When two population data are available and a confidence interval is to be build on the difference of population means the sampling statistic would be the difference of the sample means collected from these populations. If $\bar{X}_1$ and $\bar{X}_2$ are the sample means the sampling statistic would be which follows $\bar{X}_1 - \bar{X}_2$ normal distribution if both the population data are normal or approximately normal. This sampling statistic has a mean of $\mu_1 - \mu_2$ and variance of $\frac{\sigma_1^2}{n_1} + \frac{\sigma_2^2}{n_2}$. Then

$$Z = \frac{(\bar{X}_1 - \bar{X}_2) - (\mu_1 - \mu_2)}{\sqrt{\frac{\sigma_1^2}{n_1} + \frac{\sigma_2^2}{n_2}}}$$

Z follows standard normal distribution with mean 1 and variance zero.

$$P\left\{-Z\alpha/_2 \le Z \le Z\alpha/_2\right\} = 1-\alpha$$

$$P\left\{-Z\alpha/_2 \le \frac{(\bar{X}_1 - \bar{X}_2) - (\mu_1 - \mu_2)}{\sqrt{\frac{\sigma_1^2}{n_1} + \frac{\sigma_2^2}{n_2}}} \le Z\alpha/_2\right\} = 1-\alpha$$

$$P\left\{-Z\alpha/_2\sqrt{\frac{\sigma_1^2}{n_1} + \frac{\sigma_2^2}{n_2}} \le (\bar{X}_1 - \bar{X}_2) - (\mu_1 - \mu_2) \le Z\alpha/_2\sqrt{\frac{\sigma_1^2}{n_1} + \frac{\sigma_2^2}{n_2}}\right\} = 1-\alpha$$

$$P\left\{(\bar{X}_1 - \bar{X}_2) - Z\alpha/_2\sqrt{\frac{\sigma_1^2}{n_1} + \frac{\sigma_2^2}{n_2}} \le (\mu_1 - \mu_2) \le (\bar{X}_1 - \bar{X}_2) + Z\alpha/_2\sqrt{\frac{\sigma_1^2}{n_1} + \frac{\sigma_2^2}{n_2}}\right\} = 1-\alpha$$

The 100(1-α) confidence interval for $\mu_1 - \mu_2$ is as given below;

$$(\bar{X}_1 - \bar{X}_2) - Z\alpha/_2\sqrt{\frac{\sigma_1^2}{n_1} + \frac{\sigma_2^2}{n_2}} \le (\mu_1 - \mu_2) \le (\bar{X}_1 - \bar{X}_2) + Z\alpha/_2\sqrt{\frac{\sigma_1^2}{n_1} + \frac{\sigma_2^2}{n_2}}$$

Confidence Interval on Difference of two Population Means (Unknown Population Variances)

Since the variances are unknown the sample data are used in finding the estimates for the population variances. Two sub-cases may arise out of this problem. It may be assumed that either the population variances are equal or the population variances are not equal to one another.

When the Variances are not Equal

When the variances are not equal, which is reasonable to assume, the statistic $\bar{X}_1 - \bar{X}_2$ has a mean equals to $\mu_1 - \mu_2$ and variance equals to $S_1^2/n_1 + S_2^2/n_2$. Then

$$t = \frac{\left(\bar{X}_1 - \bar{X}_2\right) - \left(\mu_1 - \mu_2\right)}{\sqrt{\frac{S_1^2}{n_1} + \frac{S_2^2}{n_2}}}$$

Is approximately t-distributed with the degrees of freedom as given below;

$$\upsilon = \frac{\left(\frac{S_1^2}{n_1} - \frac{S_2^2}{n_2}\right)^2}{\frac{\left(\frac{S_1^2}{n_1}\right)^2}{n_1 + 1} + \frac{\left(\frac{S_2^2}{n_2}\right)^2}{n_1 + 1}} - 2$$

The 100 (1-α) confidence interval for $\mu_1 - \mu_2$ is as given below;

$$\left(\bar{X}_1 - \bar{X}_2\right) - t_{\alpha/2}\sqrt{\frac{S_1^2}{n_1} + \frac{S_2^2}{n_2}} \le \left(\mu_1 - \mu_2\right) \le \left(\bar{X}_1 - \bar{X}_2\right) + t_{\alpha/2}\sqrt{\frac{S_1^2}{n_1} + \frac{S_2^2}{n_2}}$$

When the Variances are Equal

When the variances are equal the sample data collected from both the populations are used together in finding the estimator for the population variance. The estimate would be as follows;

$$S_p^{\ 2} = \frac{(n_1 - 1)S_1^2 + (n_2 - 1)S_2^2}{n_1 + n_2 - 2}$$

The statistic $\bar{X}_1 - \bar{X}_2$ has a mean of $\mu_1 - \mu_2$ and the variance of $S_p^{\ 2}\left(\frac{1}{n_1} + \frac{1}{n_2}\right)$. Then

$$t_0 = \frac{\left(\bar{X}_1 - \bar{X}_2\right) - \left(\mu_1 - \mu_2\right)}{S_p\sqrt{\frac{1}{n_1} + \frac{1}{n_2}}}$$

The resulting variable follows t-distribution with $n_1 + n_2 - 2$ degrees of freedom.

The 100(1-α) confidence interval for $\mu_1 - \mu_2$ is given below;

$$\left(\bar{X}_1 - \bar{X}_2\right) - t_{\alpha/2} S_p\sqrt{\frac{1}{n_1} + \frac{1}{n_2}} \le \left(\mu_1 - \mu_2\right) \le \left(\bar{X}_1 - \bar{X}_2\right) + t_{\alpha/2} S_p\sqrt{\frac{1}{n_1} + \frac{1}{n_2}}$$

One sided intervals may also obtained by replacing α/2 with α and $t_{\alpha/2}$ with t_α.

Confidence Interval on the Population Variance

If the population is normally distributed the confidence interval on the population variance (σ^2) can be estimated using the sample data of size n and parameters $\bar{X}$, S^2, collected from the same population, and the sampling distribution of the sample variance (S^2). It can be seen that $\frac{(n-1)S^2}{\sigma^2}$ follows chi-square distribution with n-1 degrees of freedom.

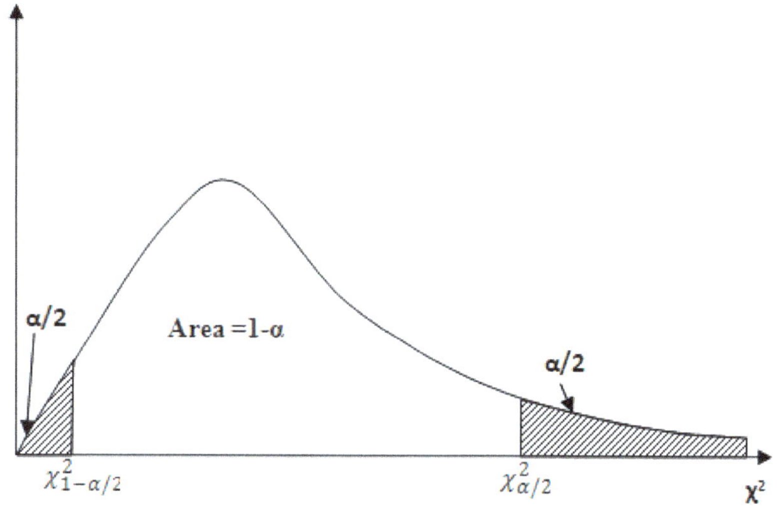

Area under the Chi-square curve corresponding to confidence level 1-α

At a particular confidence level, 1-α, the confidence interval can be estimated as follows;

$$P\left\{\chi^2_{1-\alpha/2} \le \chi^2 \le \chi^2_{\alpha/2}\right\} = 1-\alpha$$

$$P\left\{\chi^2_{1-\alpha/2} \le \frac{(n-1)S^2}{\sigma^2} \le \chi^2_{\alpha/2}\right\} = 1-\alpha$$

$$P\left\{\frac{(n-1)S^2}{\chi^2_{1-\alpha/2}} \le \sigma^2 \le \frac{(n-1)S^2}{\chi^2_{\alpha/2}}\right\} = 1-\alpha$$

The 100(1-α) confidence interval for the population variance is as follows;

$$\frac{(n-1)S^2}{\chi^2_{1-\alpha/2}} \le \sigma^2 \le \frac{(n-1)S^2}{\chi^2_{\alpha/2}}$$

The same is shown in the figure above.

Confidence Interval on the Ratios of two Population Variances

When the two populations are normally distributed with unknown means and variances, by collecting two samples from the respective populations it is possible to know the confidence interval for the ratio of the population variances. If both the samples have parameters $\bar{X}_1$, S_1^2, and $\bar{X}_2$, S_2^2 respectively, from the discussion on F-distribution it is known that $\frac{S_2^2}{\sigma_2^2} \Big/ \frac{S_1^2}{\sigma_1^2}$ follows F distribution with $n_2 - 1$ and $n_1 - 1$ degrees of freedom.

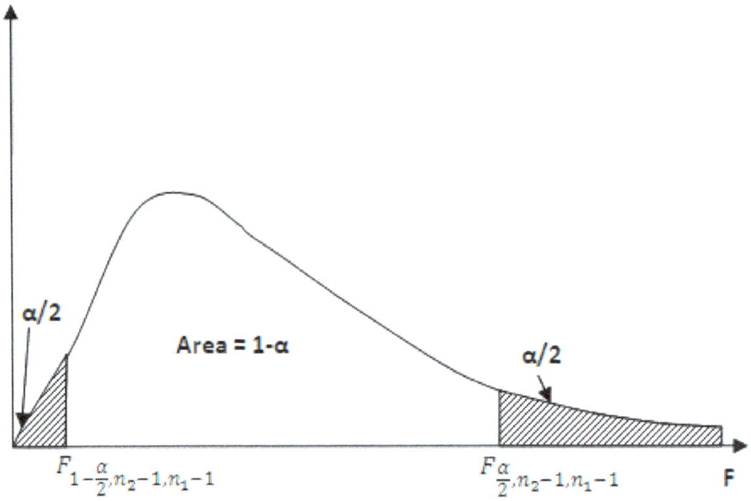

Area under the F-distribution curve corresponding to confidence level 1-α

$$P\left\{F_{1-\alpha/2} \le \frac{S_2^2}{\sigma_2^2} \Big/ \frac{S_1^2}{\sigma_1^2} \le F_{\alpha/2}\right\} = 1-\alpha$$

$$P\left\{F_{1-\alpha/2} \frac{S_1^2}{S_2^2} \le \sigma_1^2 \Big/ \sigma_2^2 \le F_{\alpha/2} \frac{S_1^2}{S_2^2}\right\} = 1-\alpha$$

The 100 (1-α) confidence interval for the ratio of the population variances is as given below;

$$F_{1-\alpha/2.n_2-1,n_1-1} \frac{S_1^2}{S_2^2} \le \sigma_1^2 \Big/ \sigma_2^2 \le F_{\alpha/2.n_2-1,n_1-1} \frac{S_1^2}{S_2^2}$$

The graphical representation of confidence interval corresponding to the significance α is shown in the figure above.

Confidence Interval on a Proportion

If the population data has certain types of data, sample data of size n collected from the population can be used in finding the confidence interval on the proportion of a particular type of data in the population. Sample›s statistic, in this case the proportion of a particular type of data $(\hat{p})$, has a mean of p and the variance of $p(1-p)/n$. Then

$$z = \frac{\hat{p} - p}{\sqrt{\frac{p(1-p)}{n}}}$$

For a given significance level α the following can be formulated;

$$P\left\{-Z_{\alpha/2} \le Z \le Z_{\alpha/2}\right\} = 1-\alpha$$

$$P\left\{-Z_{\alpha/2} \le \frac{\hat{p}-p}{\sqrt{\frac{p(1-p)}{n}}} \le Z_{\alpha/2}\right\} = 1-\alpha$$

$$P\left\{\hat{p} - Z_{\alpha/2}\sqrt{\frac{p(1-p)}{n}} \le p \le \hat{p} + Z_{\alpha/2}\sqrt{\frac{p(1-p)}{n}}\right\} = 1-\alpha$$

The 100(1-α) confidence interval for the population parameter, p, is given below;

$$\hat{p} - Z_{\alpha/2}\sqrt{\frac{p(1-p)}{n}} \le p \le \hat{p} + Z_{\alpha/2}\sqrt{\frac{p(1-p)}{n}}$$

References

- Ramsey, PH (1988). “Evaluating the Normal Approximation to the Binomial Test”. Journal of Educational Statistics. 13 (2): 173–82
- Yates, Daniel S.; David S. Moore; Daren S. Starnes (2008). The Practice of Statistics, 3rd Ed. Freeman. ISBN 978-0-7167-7309-2
- Sunter, A. B. (1977-01-01). “List Sequential Sampling with Equal or Unequal Probabilities without Replacement”. Applied Statistics. 26 (3). doi:10.2307/2346966. JSTOR 10.2307/2346966
- Bartlett, M. S.; Kendall, D. G. (1946). “The Statistical Analysis of Variance-Heterogeneity and the Logarithmic Transformation”. Supplement to the Journal of the Royal Statistical Society. 8 (1): 128–138. JSTOR 2983618
- Mood, Alexander; Graybill, Franklin A.; Boes, Duane C. (1974). Introuction to the Theory of Statistics (Third ed.). McGraw-Hill. pp. 241–246. ISBN 0-07-042864-6
- Vitter, Jeffrey S. (1985-03-01). “Random Sampling with a Reservoir”. ACM Trans. Math. Softw. 11 (1): 37–57. doi:10.1145/3147.3165. ISSN 0098-3500
- Davies, R.B. (1980). “Algorithm AS155: The Distributions of a Linear Combination of χ2 Random Variables”. Journal of the Royal Statistical Society. 29 (3): 323–333. doi:10.2307/2346911
- Jonhson, N. L.; Kotz, S.; Balakrishnan, N. (1994). “Chi-Squared Distributions including Chi and Rayleigh”. Continuous Univariate Distributions. 1 (Second ed.). John Willey and Sons. pp. 415–493. ISBN 0-471-58495-9
- Dasgupta, Sanjoy D. A.; Gupta, Anupam K. (2002). “An Elementary Proof of a Theorem of Johnson and Lindenstrauss” (PDF). Random Structures and Algorithms. 22: 60–65. doi:10.1002/rsa.10073. Retrieved 2012-05-01
- Johnson, Roger (1994), “Estimating the Size of a Population”, Teaching Statistics, 16 (2 (Summer)): 50, doi:10.1111/j.1467-9639.1994.tb00688
- Westfall, Peter H. (2013). Understanding Advanced Statistical Methods. Boca Raton, FL: CRC Press. ISBN 978-1-4665-1210-8

- Shoemaker, Lewis H. (2003). "Fixing the F Test for Equal Variances". The American Statistician. 57 (2): 105–114. doi:10.1198/0003130031441. JSTOR 30037243
- M. K. Simon, Probability Distributions Involving Gaussian Random Variables, New York: Springer, 2002, eq. (2.35), ISBN 978-0-387-34657-1

6

Statistical Hypothesis Testing: An Integrated Study

Statistical hypothesis testing helps in finding confirmatory intervals after estimates from sample data are acquired. In this method, two data, normally including a sample data, are compared to approve or disapprove a hypothesis. The chapter delves into the concept of regression analysis as well. All the diverse principles of statistical hypothesis testing have been carefully analyzed in this chapter.

Statistical Hypothesis Testing

A statistical hypothesis, sometimes called confirmatory data analysis, is a hypothesis that is testable on the basis of observing a process that is modeled via a set of random variables. A statistical hypothesis test is a method of statistical inference. Commonly, two statistical data sets are compared, or a data set obtained by sampling is compared against a synthetic data set from an idealized model. A hypothesis is proposed for the statistical relationship between the two data sets, and this is compared as an alternative to an idealized null hypothesis that proposes no relationship between two data sets. The comparison is deemed *statistically significant* if the relationship between the data sets would be an unlikely realization of the null hypothesis according to a threshold probability—the significance level. Hypothesis tests are used in determining what outcomes of a study would lead to a rejection of the null hypothesis for a pre-specified level of significance. The process of distinguishing between the null hypothesis and the alternative hypothesis is aided by identifying two conceptual types of errors (type 1 & type 2), and by specifying parametric limits on e.g. how much type 1 error will be permitted.

An alternative framework for statistical hypothesis testing is to specify a set of statistical models, one for each candidate hypothesis, and then use model selection techniques to choose the most appropriate model. The most common selection techniques are based on either Akaike information criterion or Bayes factor.

Confirmatory data analysis can be contrasted with exploratory data analysis, which may not have pre-specified hypotheses.

Variations and Sub-classes

Statistical hypothesis testing is a key technique of both frequentist inference and Bayesian inference, although the two types of inference have notable differences. Statistical hypothesis tests define a procedure that controls (fixes) the probability of incorrectly *deciding* that a default position (null hypothesis) is incorrect. The procedure is based on how likely it would be for a set of observations to occur if the null hypothesis were true. Note that this probability of making an incorrect decision is *not* the probability that the null hypothesis is true, nor whether any specific alternative hypothesis is true. This contrasts with other possible techniques of decision theory in which the null and alternative hypothesis are treated on a more equal basis.

One naïve Bayesian approach to hypothesis testing is to base decisions on the posterior probability, but this fails when comparing point and continuous hypotheses. Other approaches to decision making, such as Bayesian decision theory, attempt to balance the consequences of incorrect decisions across all possibilities, rather than concentrating on a single null hypothesis. A number of other approaches to reaching a decision based on data are available via decision theory and optimal decisions, some of which have desirable properties. Hypothesis testing, though, is a dominant approach to data analysis in many fields of science. Extensions to the theory of hypothesis testing include the study of the power of tests, i.e. the probability of correctly rejecting the null hypothesis given that it is false. Such considerations can be used for the purpose of sample size determination prior to the collection of data.

The Testing Process

In the statistics literature, statistical hypothesis testing plays a fundamental role. The usual line of reasoning is as follows:

1. There is an initial research hypothesis of which the truth is unknown.
2. The first step is to state the relevant null and alternative hypotheses. This is important, as mis-stating the hypotheses will muddy the rest of the process.
3. The second step is to consider the statistical assumptions being made about the sample in doing the test; for example, assumptions about the statistical independence or about the form of the distributions of the observations. This is equally important as invalid assumptions will mean that the results of the test are invalid.
4. Decide which test is appropriate, and state the relevant test statistic T.
5. Derive the distribution of the test statistic under the null hypothesis from the assumptions. In standard cases this will be a well-known result. For example, the test statistic might follow a Student's t distribution or a normal distribution.
6. Select a significance level (α), a probability threshold below which the null hypothesis will be rejected. Common values are 5% and 1%.
7. The distribution of the test statistic under the null hypothesis partitions the possible values of T into those for which the null hypothesis is rejected—the so-called *critical region*—and those for which it is not. The probability of the critical region is α.
8. Compute from the observations the observed value t_{obs} of the test statistic T.
9. Decide to either reject the null hypothesis in favor of the alternative or not reject it. The decision rule is to reject the null hypothesis H_0 if the observed value t_{obs} is in the critical region, and to accept or "fail to reject" the hypothesis otherwise.

An alternative process is commonly used:

1. Compute from the observations the observed value t_{obs} of the test statistic T.

2. Calculate the p-value. This is the probability, under the null hypothesis, of sampling a test statistic at least as extreme as that which was observed.

3. Reject the null hypothesis, in favor of the alternative hypothesis, if and only if the p-value is less than the significance level (the selected probability) threshold.

The two processes are equivalent. The former process was advantageous in the past when only tables of test statistics at common probability thresholds were available. It allowed a decision to be made without the calculation of a probability. It was adequate for classwork and for operational use, but it was deficient for reporting results.

The latter process relied on extensive tables or on computational support not always available. The explicit calculation of a probability is useful for reporting. The calculations are now trivially performed with appropriate software.

The difference in the two processes applied to the Radioactive suitcase example (below):

- "The Geiger-counter reading is 10. The limit is 9. Check the suitcase."
- "The Geiger-counter reading is high; 97% of safe suitcases have lower readings. The limit is 95%. Check the suitcase."

The former report is adequate, the latter gives a more detailed explanation of the data and the reason why the suitcase is being checked.

It is important to note the difference between accepting the null hypothesis and simply failing to reject it. The "fail to reject" terminology highlights the fact that the null hypothesis is assumed to be true from the start of the test; if there is a lack of evidence against it, it simply continues to be assumed true. The phrase "accept the null hypothesis" may suggest it has been proved simply because it has not been disproved, a logical fallacy known as the argument from ignorance. Unless a test with particularly high power is used, the idea of "accepting" the null hypothesis may be dangerous. Nonetheless the terminology is prevalent throughout statistics, where the meaning actually intended is well understood.

The processes described here are perfectly adequate for computation. They seriously neglect the design of experiments considerations.

It is particularly critical that appropriate sample sizes be estimated before conducting the experiment.

The phrase "test of significance" was coined by statistician Ronald Fisher.

Interpretation

If the p-value is less than the required significance level (equivalently, if the observed test statistic is in the critical region), then we say the null hypothesis is rejected at the given level of significance. Rejection of the null hypothesis is a conclusion. This is like a "guilty" verdict in a criminal trial: the evidence is sufficient to reject innocence, thus proving guilt. We might accept the alternative hypothesis (and the research hypothesis).

If the p-value is *not* less than the required significance level (equivalently, if the observed test statistic is outside the critical region), then the test has no result. The evidence is insufficient to support a conclusion. (This is like a jury that fails to reach a verdict.) The researcher typically gives extra consideration to those cases where the p-value is close to the significance level.

Some people find it helpful to think of the hypothesis testing framework as analogous to a mathematical proof by contradiction.

In the Lady tasting tea example (below), Fisher required the Lady to properly categorize all of the cups of tea to justify the conclusion that the result was unlikely to result from chance. He defined the critical region as that case alone. The region was defined by a probability (that the null hypothesis was correct) of less than 5%.

Whether rejection of the null hypothesis truly justifies acceptance of the research hypothesis depends on the structure of the hypotheses. Rejecting the hypothesis that a large paw print originated from a bear does not immediately prove the existence of Bigfoot. Hypothesis testing emphasizes the rejection, which is based on a probability, rather than the acceptance, which requires extra steps of logic.

"The probability of rejecting the null hypothesis is a function of five factors: whether the test is one- or two tailed, the level of significance, the standard deviation, the amount of deviation from the null hypothesis, and the number of observations." These factors are a source of criticism; factors under the control of the experimenter/analyst give the results an appearance of subjectivity.

Use and Importance

Statistics are helpful in analyzing most collections of data. This is equally true of hypothesis testing which can justify conclusions even when no scientific theory exists. In the Lady tasting tea example, it was "obvious" that no difference existed between (milk poured into tea) and (tea poured into milk). The data contradicted the "obvious".

Real world applications of hypothesis testing include:

- Testing whether more men than women suffer from nightmares
- Establishing authorship of documents
- Evaluating the effect of the full moon on behavior
- Determining the range at which a bat can detect an insect by echo
- Deciding whether hospital carpeting results in more infections
- Selecting the best means to stop smoking
- Checking whether bumper stickers reflect car owner behavior
- Testing the claims of handwriting analysts

Statistical hypothesis testing plays an important role in the whole of statistics and in statistical inference. For example, Lehmann (1992) in a review of the fundamental paper by Neyman and Pearson (1933) says: "Nevertheless, despite their shortcomings, the new paradigm formulated in the 1933 paper, and the many developments carried out within its framework continue to play a central role in both the theory and practice of statistics and can be expected to do so in the foreseeable future".

Significance testing has been the favored statistical tool in some experimental social sciences. Other fields have favored the estimation of parameters (e.g., effect size). Significance testing is used as a substitute for the traditional comparison of predicted value and experimental result at the core of the scientific method. When theory is only capable of predicting the sign of a relationship, a directional (one-sided) hypothesis test can be configured so that only a statistically significant result supports theory. This form of theory appraisal is the most heavily criticized application of hypothesis testing.

Cautions

"If the government required statistical procedures to carry warning labels like those on drugs, most inference methods would have long labels indeed." This caution applies to hypothesis tests and alternatives to them.

The successful hypothesis test is associated with a probability and a type-I error rate. The conclusion *might* be wrong.

The conclusion of the test is only as solid as the sample upon which it is based. The design of the experiment is critical. A number of unexpected effects have been observed including:

- The clever Hans effect. A horse appeared to be capable of doing simple arithmetic.
- The Hawthorne effect. Industrial workers were more productive in better illumination, and most productive in worse.
- The placebo effect. Pills with no medically active ingredients were remarkably effective.

A statistical analysis of misleading data produces misleading conclusions. The issue of data quality can be more subtle. In forecasting for example, there is no agreement on a measure of forecast accuracy. In the absence of a consensus measurement, no decision based on measurements will be without controversy.

The book *How to Lie with Statistics* is the most popular book on statistics ever published. It does not much consider hypothesis testing, but its cautions are applicable, including: Many claims are made on the basis of samples too small to convince. If a report does not mention sample size, be doubtful.

Hypothesis testing acts as a filter of statistical conclusions; only those results meeting a probability threshold are publishable. Economics also acts as a publication filter; only those results favorable to the author and funding source may be submitted for publication. The impact of filtering on publication is termed publication bias. A related problem is that of multiple testing (sometimes linked to data mining), in which a variety of tests for a variety of possible effects are applied to a single data set and only those yielding a significant result are reported. These are often dealt with by using multiplicity correction procedures that control the family wise error rate (FWER) or the false discovery rate (FDR).

Those making critical decisions based on the results of a hypothesis test are prudent to look at the details rather than the conclusion alone. In the physical sciences most results are fully accepted only when independently confirmed. The general advice concerning statistics is, "Figures never lie, but liars figure" (anonymous).

Examples

Lady Tasting Tea

In a famous example of hypothesis testing, known as the *Lady tasting tea*, Dr. Muriel Bristol, a female colleague of Fisher claimed to be able to tell whether the tea or the milk was added first to a cup. Fisher proposed to give her eight cups, four of each variety, in random order. One could then ask what the probability was for her getting the number she got correct, but just by chance. The null hypothesis was that the Lady had no such ability. The test statistic was a simple count of the number of successes in selecting the 4 cups. The critical region was the single case of 4 successes of 4 possible based on a conventional probability criterion (< 5%; 1 of 70 ≈ 1.4%). Fisher asserted that no alternative hypothesis was (ever) required. The lady correctly identified every cup, which would be considered a statistically significant result.

Courtroom Trial

A statistical test procedure is comparable to a criminal trial; a defendant is considered not guilty as long as his or her guilt is not proven. The prosecutor tries to prove the guilt of the defendant. Only when there is enough charging evidence the defendant is convicted.

In the start of the procedure, there are two hypotheses H_0: "the defendant is not guilty", and H_1: "the defendant is guilty". The first one, H_0, is called the *null hypothesis*, and is for the time being accepted. The second one, H_1, is called the *alternative hypothesis*. It is the alternative hypothesis that one hopes to support.

The hypothesis of innocence is only rejected when an error is very unlikely, because one doesn't want to convict an innocent defendant. Such an error is called *error of the first kind* (i.e., the conviction of an innocent person), and the occurrence of this error is controlled to be rare. As a consequence of this asymmetric behaviour, the *error of the second kind* (acquitting a person who committed the crime), is often rather large.

	H_0 is true Truly not guilty	**H_1 is true Truly guilty**
Accept null hypothesis Acquittal	Right decision	Wrong decision Type II Error
Reject null hypothesis Conviction	Wrong decision Type I Error	Right decision

A criminal trial can be regarded as either or both of two decision processes: guilty vs not guilty or evidence vs a threshold ("beyond a reasonable doubt"). In one view, the defendant is judged; in the other view the performance of the prosecution (which bears the burden of proof) is judged. A hypothesis test can be regarded as either a judgment of a hypothesis or as a judgment of evidence.

Philosopher's Beans

The following example was produced by a philosopher describing scientific methods generations before hypothesis testing was formalized and popularized.

Few beans of this handful are white.
Most beans in this bag are white.
Therefore: Probably, these beans were taken from another bag.
This is an hypothetical inference.

The beans in the bag are the population. The handful are the sample. The null hypothesis is that the sample originated from the population. The criterion for rejecting the null-hypothesis is the "obvious" difference in appearance (an informal difference in the mean). The interesting result is that consideration of a real population and a real sample produced an imaginary bag. The philosopher was considering logic rather than probability. To be a real statistical hypothesis test, this example requires the formalities of a probability calculation and a comparison of that probability to a standard.

A simple generalization of the example considers a mixed bag of beans and a handful that contain either very few or very many white beans. The generalization considers both extremes. It requires more calculations and more comparisons to arrive at a formal answer, but the core philosophy is unchanged; If the composition of the handful is greatly different from that of the bag, then the sample probably originated from another bag. The original example is termed a one-sided or a one-tailed test while the generalization is termed a two-sided or two-tailed test.

The statement also relies on the inference that the sampling was random. If someone had been picking through the bag to find white beans, then it would explain why the handful had so many white beans, and also explain why the number of white beans in the bag was depleted (although the bag is probably intended to be assumed much larger than one's hand).

Clairvoyant Card Game

A person (the subject) is tested for clairvoyance. He is shown the reverse of a randomly chosen playing card 25 times and asked which of the four suits it belongs to. The number of hits, or correct answers, is called X.

As we try to find evidence of his clairvoyance, for the time being the null hypothesis is that the person is not clairvoyant. The alternative is, of course: the person is (more or less) clairvoyant.

If the null hypothesis is valid, the only thing the test person can do is guess. For every card, the probability (relative frequency) of any single suit appearing is 1/4. If the alternative is valid, the test subject will predict the suit correctly with probability greater than 1/4. We will call the probability of guessing correctly p. The hypotheses, then, are:

- null hypothesis : $H_0 : p = \frac{1}{4}$ (just guessing)

and

- alternative hypothesis :$H_1 : p > \frac{1}{4}$ (true clairvoyant).

When the test subject correctly predicts all 25 cards, we will consider him clairvoyant, and reject the null hypothesis. Thus also with 24 or 23 hits. With only 5 or 6 hits, on the other hand, there is no cause to consider him so. But what about 12 hits, or 17 hits? What is the critical number, c, of hits, at which point we consider the subject to be clairvoyant? How do we determine the critical value c? It is obvious that with the choice c=25 (i.e. we only accept clairvoyance when all cards are predicted correctly) we're more critical than with c=10. In the first case almost no test subjects will be recognized to be clairvoyant, in the second case, a certain number will pass the test. In practice, one decides how critical one will be. That is, one decides how often one accepts an error of the first kind – a false positive, or Type I error. With c = 25 the probability of such an error is:

$$P(\text{reject } H_0 \mid H_0 \text{ is valid}) = P(X = 25 \mid p = \tfrac{1}{4}) = \left(\tfrac{1}{4}\right)^{25} \approx 10^{-15},$$

and hence, very small. The probability of a false positive is the probability of randomly guessing correctly all 25 times.

Being less critical, with c=10, gives:

$$P(\text{reject } H_0 \mid H_0 \text{ is valid}) = P(X \geq 10 \mid p = \tfrac{1}{4}) = \sum_{k=10}^{25} P(X = k \mid p = \tfrac{1}{4}) \approx 0.07.$$

Thus, c = 10 yields a much greater probability of false positive.

Before the test is actually performed, the maximum acceptable probability of a Type I error (α) is determined. Typically, values in the range of 1% to 5% are selected. (If the maximum acceptable error rate is zero, an infinite number of correct guesses is required.) Depending on this Type 1 error rate, the critical value c is calculated. For example, if we select an error rate of 1%, c is calculated thus:

$$P(\text{reject } H_0 \mid H_0 \text{ is valid}) = P(X \geq c \mid p = \tfrac{1}{4}) \leq 0.01.$$

From all the numbers c, with this property, we choose the smallest, in order to minimize the probability of a Type II error, a false negative. For the above example, we select: $c = 13$.

Radioactive Suitcase

As an example, consider determining whether a suitcase contains some radioactive material. Placed under a Geiger counter, it produces 10 counts per minute. The null hypothesis is that no radioactive material is in the suitcase and that all measured counts are due to ambient radioactivity typical of the surrounding air and harmless objects. We can then calculate how likely it is that we would observe 10 counts per minute if the null hypothesis were true. If the null hypothesis predicts (say) on average 9 counts per minute, then according to the Poisson distribution typical for radioactive decay there is about 41% chance of recording 10 or more counts. Thus we can say that the suitcase is compatible with the null hypothesis (this does not guarantee that there is no radioactive material, just that we don't have enough evidence to suggest there is). On the other hand, if the null hypothesis predicts 3 counts per minute (for which the Poisson distribution predicts only 0.1% chance of recording 10 or more counts) then the suitcase is not compatible with the null hypothesis, and there are likely other factors responsible to produce the measurements.

The test does not directly assert the presence of radioactive material. A *successful* test asserts that the claim of no radioactive material present is unlikely given the reading (and therefore ...). The double negative (disproving the null hypothesis) of the method is confusing, but using a counter-example to disprove is standard mathematical practice. The attraction of the method is its practicality. We know (from experience) the expected range of counts with only ambient radioactivity present, so we can say that a measurement is *unusually* large. Statistics just formalizes the intuitive by using numbers instead of adjectives. We probably do not know the characteristics of the radioactive suitcases; We just assume that they produce larger readings.

To slightly formalize intuition: Radioactivity is suspected if the Geiger-count with the suitcase is among or exceeds the greatest (5% or 1%) of the Geiger-counts made with ambient radiation alone. This makes no assumptions about the distribution of counts. Many ambient radiation observations are required to obtain good probability estimates for rare events.

The test described here is more fully the null-hypothesis statistical significance test. The null hypothesis represents what we would believe by default, before seeing any evidence. Statistical significance is a possible finding of the test, declared when the observed sample is unlikely to have occurred by chance if the null hypothesis were true. The name of the test describes its formulation and its possible outcome. One characteristic of the test is its crisp decision: to reject or not reject the null hypothesis. A calculated value is compared to a threshold, which is determined from the tolerable risk of error.

Definition of Terms

The following definitions are mainly based on the exposition in the book by Lehmann and Romano:

Statistical hypothesis

A statement about the parameters describing a population (not a sample).

Statistic

A value calculated from a sample, often to summarize the sample for comparison purposes.

Simple hypothesis

Any hypothesis which specifies the population distribution completely.

Composite hypothesis

Any hypothesis which does *not* specify the population distribution completely.

Null hypothesis (H_0)

A simple hypothesis associated with a contradiction to a theory one would like to prove.

Alternative hypothesis (H_1)

A hypothesis (often composite) associated with a theory one would like to prove.

Statistical test

A procedure whose inputs are samples and whose result is a hypothesis.

Region of acceptance

The set of values of the test statistic for which we fail to reject the null hypothesis.

Region of rejection / Critical region

The set of values of the test statistic for which the null hypothesis is rejected.

Critical value

The threshold value delimiting the regions of acceptance and rejection for the test statistic.

Power of a test $(1 - \beta)$

The test's probability of correctly rejecting the null hypothesis. The complement of the false negative rate, β. Power is termed sensitivity in biostatistics. ("This is a sensitive test. Because the result is negative, we can confidently say that the patient does not have the condition.")

Size

For simple hypotheses, this is the test's probability of *incorrectly* rejecting the null hypothesis. The false positive rate. For composite hypotheses this is the supremum of the probability of rejecting the null hypothesis over all cases covered by the null hypothesis. The complement of the false positive rate is termed specificity in biostatistics. ("This is a specific test. Because the result is positive, we can confidently say that the patient has the condition.")

Significance level of a test (α)

It is the upper bound imposed on the size of a test. Its value is chosen by the statistician prior to looking at the data or choosing any particular test to be used. It is the maximum exposure to erroneously rejecting H_0 he/she is ready to accept. Testing H_0 at significance level α means testing H_0 with a test whose size does not exceed α. In most cases, one uses tests whose size is equal to the significance level.

p-value

The probability, assuming the null hypothesis is true, of observing a result at least as extreme as the test statistic.

Statistical significance test

A predecessor to the statistical hypothesis test. An experimental result was said to be statistically significant if a sample was sufficiently inconsistent with the (null) hypothesis. This was variously considered common sense, a pragmatic heuristic for identifying meaningful experimental results, a convention establishing a threshold of statistical evidence or a method for drawing conclusions from data. The statistical hypothesis test added mathematical rigor and philosophical consistency to the concept by making the alternative hypothesis explicit. The term is loosely used to describe the modern version which is now part of statistical hypothesis testing.

Conservative test

A test is conservative if, when constructed for a given nominal significance level, the true probability of *incorrectly* rejecting the null hypothesis is never greater than the nominal level.

Exact test

A test in which the significance level or critical value can be computed exactly, i.e., without any approximation. In some contexts this term is restricted to tests applied to categorical data and to permutation tests, in which computations are carried out by complete enumeration of all possible outcomes and their probabilities.

A statistical hypothesis test compares a test statistic (z or t for examples) to a threshold. The test statistic (the formula found in the table below) is based on optimality. For a fixed level of Type I error rate, use of these statistics minimizes Type II error rates (equivalent to maximizing power). The following terms describe tests in terms of such optimality:

Most powerful test

For a given *size* or *significance level*, the test with the greatest power (probability of rejection) for a given value of the parameter(s) being tested, contained in the alternative hypothesis.

Uniformly most powerful test (UMP)

A test with the greatest *power* for all values of the parameter(s) being tested, contained in the alternative hypothesis.

Common Test Statistics

One-sample tests are appropriate when a sample is being compared to the population from a hypothesis. The population characteristics are known from theory or are calculated from the population.

Two-sample tests are appropriate for comparing two samples, typically experimental and control samples from a scientifically controlled experiment.

Paired tests are appropriate for comparing two samples where it is impossible to control important variables. Rather than comparing two sets, members are paired between samples so the difference between the members becomes the sample. Typically the mean of the differences is then compared to zero. The common example scenario for when a paired difference test is appropriate is when a single set of test subjects has something applied to them and the test is intended to check for an effect.

Z-tests are appropriate for comparing means under stringent conditions regarding normality and a known standard deviation.

A *t*-test is appropriate for comparing means under relaxed conditions (less is assumed).

Tests of proportions are analogous to tests of means (the 50% proportion).

Chi-squared tests use the same calculations and the same probability distribution for different applications:

- Chi-squared tests for variance are used to determine whether a normal population has a specified variance. The null hypothesis is that it does.
- Chi-squared tests of independence are used for deciding whether two variables are associated or are independent. The variables are categorical rather than numeric. It can be used to decide whether left-handedness is correlated with libertarian politics (or not). The null hypothesis is that the variables are independent. The numbers used in the calculation are the observed and expected frequencies of occurrence (from contingency tables).
- Chi-squared goodness of fit tests are used to determine the adequacy of curves fit to data. The null hypothesis is that the curve fit is adequate. It is common to determine curve shapes to minimize the mean square error, so it is appropriate that the goodness-of-fit calculation sums the squared errors.

F-tests (analysis of variance, ANOVA) are commonly used when deciding whether groupings of data by category are meaningful. If the variance of test scores of the left-handed in a class is much smaller than the variance of the whole class, then it may be useful to study lefties as a group. The null hypothesis is that two variances are the same – so the proposed grouping is not meaningful.

In the table below, the symbols used are defined at the bottom of the table. Proofs exist that the test statistics are appropriate.

Name	Formula	Assumptions or notes
One-sample z-test	$z = \dfrac{\bar{x} - \mu_0}{(\sigma / \sqrt{n})}$	(Normal population or $n > 30$) and σ known. (z is the distance from the mean in relation to the standard deviation of the mean). For non-normal distributions it is possible to calculate a minimum proportion of a population that falls within k standard deviations for any k.
Two-sample z-test	$z = \dfrac{(\bar{x}_1 - \bar{x}_2) - d_0}{\sqrt{\dfrac{\sigma_1^2}{n_1} + \dfrac{\sigma_2^2}{n_2}}}$	Normal population and independent observations and σ_1 and σ_2 are known
One-sample t-test	$t = \dfrac{\bar{x} - \mu_0}{(s / \sqrt{n})}$, $df = n - 1$	(Normal population or $n > 30$) and σ unknown
Paired t-test	$t = \dfrac{\bar{d} - d_0}{(s_d / \sqrt{n})}$, $df = n - 1$	(Normal population of differences or $n >$ 30) and σ unknown or small sample size $n < 30$

Two-sample pooled *t*-test, equal variances	$t=\frac{(\bar{x}_1-\bar{x}_2)-d_0}{s_p\sqrt{\frac{1}{n_1}+\frac{1}{n_2}}}$, $s_p^2=\frac{(n_1-1)s_1^2+(n_2-1)s_2^2}{n_1+n_2-2}$, $df=n_1+n_2-2$	(Normal populations or $n_1+n_2>40$) and independent observations and $\sigma_1=\sigma_2$ unknown
Two-sample unpooled *t*-test, unequal variances (Welch's *t*-test)	$t=\frac{(\bar{x}_1-\bar{x}_2)-d_0}{\sqrt{\frac{s_1^2}{n_1}+\frac{s_2^2}{n_2}}}$, $df=\frac{\left(\frac{s_1^2}{n_1}+\frac{s_2^2}{n_2}\right)^2}{\frac{\left(\frac{s_1^2}{n_1}\right)^2}{n_1-1}+\frac{\left(\frac{s_2^2}{n_2}\right)^2}{n_2-1}}$	(Normal populations or $n_1+n_2>40$) and independent observations and $\sigma_1\neq\sigma_2$ both unknown
One-proportion z-test	$z=\frac{\hat{p}-p_0}{\sqrt{p_0(1-p_0)}}\sqrt{n}$	$n\,p_0>10$ and $n\,(1-p_0)>10$ and it is a SRS (Simple Random Sample).
Two-proportion z-test, pooled for $H_0: p_1=p_2$	$z=\frac{(\hat{p}_1-\hat{p}_2)}{\sqrt{\hat{p}(1-\hat{p})(\frac{1}{n_1}+\frac{1}{n_2})}}$, $\hat{p}=\frac{x_1+x_2}{n_1+n_2}$	$n_1\,p_1>5$ and $n_1(1-p_1)>5$ and $n_2\,p_2>5$ and $n_2(1-p_2)>5$ and independent observations.
Two-proportion z-test, unpooled for $\|d_0\|>0$	$z=\frac{(\hat{p}_1-\hat{p}_2)-d_0}{\sqrt{\frac{\hat{p}_1(1-\hat{p}_1)}{n_1}+\frac{\hat{p}_2(1-\hat{p}_2)}{n_2}}}$	$n_1\,p_1>5$ and $n_1(1-p_1)>5$ and $n_2\,p_2>5$ and $n_2(1-p_2)>5$ and independent observations.
Chi-squared test for variance	$\chi^2=(n-1)\frac{s^2}{\sigma_0^2}$	Normal population

Chi-squared test for goodness of fit	$\chi^2 = \sum^{k} \frac{(\text{observed} - \text{expected})^2}{\text{expected}}$	*df = k* – 1 – *# parameters estimated*, and one of these must hold. • All expected counts are at least 5. • All expected counts are > 1 and no more than 20% of expected counts are less than 5
Two-sample F test for equality of variances	$F = \frac{s_1^2}{s_2^2}$	Normal populations Arrange so $s_1^2 \geq s_2^2$ and reject H_0 for $F > F(\alpha/2, n_1 - 1, n_2 - 1)$
Regression *t*-test of $H_0 : R^2 = 0.$	$t = \sqrt{\frac{R^2(n-k-1^*)}{1-R^2}}$	Reject H_0 for $t > t(\alpha/2, n-k-1^*)$ *Subtract 1 for intercept; *k* terms contain independent variables.

In general, the subscript 0 indicates a value taken from the null hypothesis, H_0, which should be used as much as possible in constructing its test statistic. ... *Definitions of other symbols:*

- α, the probability of Type I error (rejecting a null hypothesis when it is in fact true)
- n = sample size
- n_1 = sample 1 size
- n_2 = sample 2 size
- $\bar{x}$ = sample mean
- μ_0 = hypothesized population mean
- μ_1 = population 1 mean
- μ_2 = population 2 mean
- σ = population standard deviation
- σ^2 = population variance
- s = sample standard deviation
- $\sum^{k}$ = sum (of *k* numbers)

- s^2 = sample variance
- s_1 = sample 1 standard deviation
- s_2 = sample 2 standard deviation
- t = t statistic
- df = degrees of freedom
- $\bar{d}$ = sample mean of differences
- d_0 = hypothesized population mean difference
- s_d = standard deviation of differences
- χ^2 = Chi-squared statistic

- $\hat{p} = x/n$ = sample proportion, unless specified otherwise
- p_0 = hypothesized population proportion
- p_1 = proportion 1
- p_2 = proportion 2
- d_p = hypothesized difference in proportion
- $\min\{n_1, n_2\}$ = minimum of n_1 and n_2
- $x_1 = n_1 p_1$
- $x_2 = n_2 p_2$
- F = F statistic

Origins and Early Controversy

Significance testing is largely the product of Karl Pearson (p-value, Pearson's chi-squared test), William Sealy Gosset (Student's t-distribution), and Ronald Fisher ("null hypothesis", analysis of variance, "significance test"), while hypothesis testing was developed by Jerzy Neyman and Egon Pearson (son of Karl). Ronald Fisher began his life in statistics as a Bayesian (Zabell 1992), but Fisher soon grew disenchanted with the subjectivity involved (namely use of the principle of indifference when determining prior probabilities), and sought to provide a more "objective" approach to inductive inference.

Fisher was an agricultural statistician who emphasized rigorous experimental design and methods to extract a result from few samples assuming Gaussian distributions. Neyman (who teamed with the younger Pearson) emphasized mathematical rigor and methods to obtain more results from many samples and a wider range of distributions. Modern hypothesis testing is an inconsistent hybrid of the Fisher vs Neyman/Pearson formulation, methods and terminology developed in the early 20th century. While hypothesis testing was popularized early in the 20th century, evidence of its use can be found much earlier. In the 1770s Laplace considered the statistics of almost half a million births. The statistics showed an excess of boys compared to girls. He concluded by calculation of a p-value that the excess was a real, but unexplained, effect.

Fisher popularized the "significance test". He required a null-hypothesis (corresponding to a population frequency distribution) and a sample. His (now familiar) calculations determined whether to reject the null-hypothesis or not. Significance testing did not utilize an alternative hypothesis so there was no concept of a Type II error.

The p-value was devised as an informal, but objective, index meant to help a researcher determine (based on other knowledge) whether to modify future experiments or strengthen one's faith in the null hypothesis. Hypothesis testing (and Type I/II errors) was devised by Neyman and Pearson as a more objective alternative to Fisher's p-value, also meant to determine researcher behaviour, but without requiring any inductive inference by the researcher.

Neyman & Pearson considered a different problem (which they called "hypothesis testing"). They initially considered two simple hypotheses (both with frequency distributions). They calculated two probabilities and typically selected the hypothesis associated with the higher probability (the hypothesis more likely to have generated the sample). Their method always selected a hypothesis. It also allowed the calculation of both types of error probabilities.

Fisher and Neyman/Pearson clashed bitterly. Neyman/Pearson considered their formulation to be an improved generalization of significance testing.(The defining paper was abstract. Mathematicians have generalized and refined the theory for decades.) Fisher thought that it was not applicable to scientific research because often, during the course of the experiment, it is discovered that the initial assumptions about the null hypothesis are questionable due to unexpected sources of error. He believed that the use of rigid reject/accept decisions based on models formulated before data is collected was incompatible with this common scenario faced by scientists and attempts to apply this method to scientific research would lead to mass confusion.

The dispute between Fisher and Neyman–Pearson was waged on philosophical grounds, characterized by a philosopher as a dispute over the proper role of models in statistical inference.

Events intervened: Neyman accepted a position in the western hemisphere, breaking his partnership with Pearson and separating disputants (who had occupied the same building) by much of the planetary diameter. World War II provided an intermission in the debate. The dispute between Fisher and Neyman terminated (unresolved after 27 years) with Fisher's death in 1962. Neyman wrote a well-regarded eulogy. Some of Neyman's later publications reported p-values and significance levels.

The modern version of hypothesis testing is a hybrid of the two approaches that resulted from confusion by writers of statistical textbooks (as predicted by Fisher) beginning in the 1940s. (But signal detection, for example, still uses the Neyman/Pearson formulation.) Great conceptual differences and many caveats in addition to those mentioned above were ignored. Neyman and Pearson provided the stronger terminology, the more rigorous mathematics and the more consistent philosophy, but the subject taught today in introductory statistics has more similarities with Fisher's method than theirs. This history explains the inconsistent terminology (example: the null hypothesis is never accepted, but there is a region of acceptance).

Sometime around 1940, in an apparent effort to provide researchers with a "non-controversial" way to have their cake and eat it too, the authors of statistical text books began anonymously combining these two strategies by using the p-value in place of the test statistic (or data) to test against the Neyman–Pearson "significance level". Thus, researchers were encouraged to infer the strength of their data against some null hypothesis using p-values, while also thinking they are retaining the post-data collection objectivity provided by hypothesis testing. It then became customary for the null hypothesis, which was originally some realistic research hypothesis, to be used almost solely as a strawman "nil" hypothesis (one where a treatment has no effect, regardless of the context).

A comparison between Fisherian, Frequentist (Neyman–Pearson)

Fisher's null hypothesis testing	Neyman–Pearson decision theory
1. Set up a statistical null hypothesis. The null need not be a nil hypothesis (i.e., zero difference).	1. Set up two statistical hypotheses, H1 and H2, and decide about α, β, and sample size before the experiment, based on subjective cost-benefit considerations. These define a rejection region for each hypothesis.
2. Report the exact level of significance (e.g., $p = 0.051$ or $p = 0.049$). Do not use a conventional 5% level, and do not talk about accepting or rejecting hypotheses. If the result is "not significant", draw no conclusions and make no decisions, but suspend judgement until further data is available.	2. If the data falls into the rejection region of H1, accept H2; otherwise accept H1. Note that accepting a hypothesis does not mean that you believe in it, but only that you act as if it were true.
3. Use this procedure only if little is known about the problem at hand, and only to draw provisional conclusions in the context of an attempt to understand the experimental situation.	3. The usefulness of the procedure is limited among others to situations where you have a disjunction of hypotheses (e.g., either $\mu 1 = 8$ or $\mu 2 = 10$ is true) and where you can make meaningful cost-benefit trade-offs for choosing alpha and beta.

Early Choices of Null Hypothesis

Paul Meehl has argued that the epistemological importance of the choice of null hypothesis has gone largely unacknowledged. When the null hypothesis is predicted by theory, a more precise experiment will be a more severe test of the underlying theory. When the null hypothesis defaults to "no difference" or "no effect", a more precise experiment is a less severe test of the theory that motivated performing the experiment. An examination of the origins of the latter practice may therefore be useful:

1778: Pierre Laplace compares the birthrates of boys and girls in multiple European cities. He states: "it is natural to conclude that these possibilities are very nearly in the same ratio". Thus Laplace's null hypothesis that the birthrates of boys and girls should be equal given "conventional wisdom".

1900: Karl Pearson develops the chi squared test to determine "whether a given form of frequency curve will effectively describe the samples drawn from a given population." Thus the null hypothesis is that a population is described by some distribution predicted by theory. He uses as an example the numbers of five and sixes in the Weldon dice throw data.

1904: Karl Pearson develops the concept of "contingency" in order to determine whether outcomes are independent of a given categorical factor. Here the null hypothesis is by default that two things are unrelated (e.g. scar formation and death rates from smallpox). The null hypothesis in this case is no longer predicted by theory or conventional wisdom, but is instead the principle of indifference that lead Fisher and others to dismiss the use of "inverse probabilities".

Null Hypothesis Statistical Significance Testing

An example of Neyman–Pearson hypothesis testing can be made by a change to the radioactive suitcase example. If the "suitcase" is actually a shielded container for the transportation of radioactive material, then a test might be used to select among three hypotheses: no radioactive source present, one present, two (all) present. The test could be required for safety, with actions required in each case. The Neyman–Pearson lemma of hypothesis testing says that a good criterion for the selection of hypotheses is the ratio of their probabilities (a likelihood ratio). A simple method of solution is to select the hypothesis with the highest probability for the Geiger counts observed. The typical result matches intuition: few counts imply no source, many counts imply two sources and intermediate counts imply one source.

Neyman–Pearson theory can accommodate both prior probabilities and the costs of actions resulting from decisions. The former allows each test to consider the results of earlier tests (unlike Fisher's significance tests). The latter allows the consideration of economic issues (for example) as well as probabilities. A likelihood ratio remains a good criterion for selecting among hypotheses.

The two forms of hypothesis testing are based on different problem formulations. The original test is analogous to a true/false question; the Neyman–Pearson test is more like multiple choice. In the view of Tukey the former produces a conclusion on the basis of only strong evidence while the latter produces a decision on the basis of available evidence. While the two tests seem quite different both mathematically and philosophically, later developments lead to the opposite claim. Consider many tiny radioactive sources. The hypotheses become 0,1,2,3... grains of radioactive sand. There is little distinction between none or some radiation (Fisher) and 0 grains of radioactive sand versus all of the alternatives (Neyman–Pearson). The major Neyman–Pearson paper of 1933 also considered composite hypotheses (ones whose distribution includes an unknown parameter). An example proved the optimality of the (Student's) *t*-test, "there can be no better test for the hypothesis under consideration" (p 321). Neyman–Pearson theory was proving the optimality of Fisherian methods from its inception.

Fisher's significance testing has proven a popular flexible statistical tool in application with little mathematical growth potential. Neyman–Pearson hypothesis testing is claimed as a pillar of mathematical statistics, creating a new paradigm for the field. It also stimulated new applications in statistical process control, detection theory, decision theory and game theory. Both formulations have been successful, but the successes have been of a different character.

The dispute over formulations is unresolved. Science primarily uses Fisher's (slightly modified) formulation as taught in introductory statistics. Statisticians study Neyman–Pearson theory in graduate school. Mathematicians are proud of uniting the formulations. Philosophers consider them separately. Learned opinions deem the formulations variously competitive (Fisher vs Neyman), incompatible or complementary. The dispute has become more complex since Bayesian inference has achieved respectability.

The terminology is inconsistent. Hypothesis testing can mean any mixture of two formulations that both changed with time. Any discussion of significance testing vs hypothesis testing is doubly vulnerable to confusion.

Fisher thought that hypothesis testing was a useful strategy for performing industrial quality control, however, he strongly disagreed that hypothesis testing could be useful for scientists. Hypothesis testing provides a means of finding test statistics used in significance testing. The concept of power is useful in explaining the consequences of adjusting the significance level and is heavily used in sample size determination. The two methods remain philosophically distinct. They usually (but *not always*) produce the same mathematical answer. The preferred answer is context dependent. While the existing merger of Fisher and Neyman–Pearson theories has been heavily criticized, modifying the merger to achieve Bayesian goals has been considered.

Criticism

Criticism of statistical hypothesis testing fills volumes citing 300–400 primary references. Much of the criticism can be summarized by the following issues:

- The interpretation of a p-value is dependent upon stopping rule and definition of multiple comparison. The former often changes during the course of a study and the latter is unavoidably ambiguous. (i.e. "p values depend on both the (data) observed and on the other possible (data) that might have been observed but weren't").
- Confusion resulting (in part) from combining the methods of Fisher and Neyman–Pearson which are conceptually distinct.
- Emphasis on statistical significance to the exclusion of estimation and confirmation by repeated experiments.
- Rigidly requiring statistical significance as a criterion for publication, resulting in publication bias. Most of the criticism is indirect. Rather than being wrong, statistical hypothesis testing is misunderstood, overused and misused.

- When used to detect whether a difference exists between groups, a paradox arises. As improvements are made to experimental design (e.g., increased precision of measurement and sample size), the test becomes more lenient. Unless one accepts the absurd assumption that all sources of noise in the data cancel out completely, the chance of finding statistical significance in either direction approaches 100%.
- Layers of philosophical concerns. The probability of statistical significance is a function of decisions made by experimenters/analysts. If the decisions are based on convention they are termed arbitrary or mindless while those not so based may be termed subjective. To minimize type II errors, large samples are recommended. In psychology practically all null hypotheses are claimed to be false for sufficiently large samples so "...it is usually nonsensical to perform an experiment with the *sole* aim of rejecting the null hypothesis.". "Statistically significant findings are often misleading" in psychology. Statistical significance does not imply practical significance and correlation does not imply causation. Casting doubt on the null hypothesis is thus far from directly supporting the research hypothesis.
- "[I]t does not tell us what we want to know". Lists of dozens of complaints are available.

Critics and supporters are largely in factual agreement regarding the characteristics of null hypothesis significance testing (NHST): While it can provide critical information, it is *inadequate as the sole tool for statistical analysis. Successfully rejecting the null hypothesis may offer no support for the research hypothesis.* The continuing controversy concerns the selection of the best statistical practices for the near-term future given the (often poor) existing practices. Critics would prefer to ban NHST completely, forcing a complete departure from those practices, while supporters suggest a less absolute change.

Controversy over significance testing, and its effects on publication bias in particular, has produced several results. The American Psychological Association has strengthened its statistical reporting requirements after review, medical journal publishers have recognized the obligation to publish some results that are not statistically significant to combat publication bias and a journal has been created to publish such results exclusively. Textbooks have added some cautions and increased coverage of the tools necessary to estimate the size of the sample required to produce significant results. Major organizations have not abandoned use of significance tests although some have discussed doing so.

Alternatives

The numerous criticisms of significance testing do not lead to a single alternative. A unifying position of critics is that statistics should not lead to a conclusion or a decision but to a probability or to an estimated value with a confidence interval rather than to an accept-reject decision regarding a particular hypothesis. It is unlikely that the controversy surrounding significance testing will be resolved in the near future. Its supposed flaws and unpopularity do not eliminate the need for an objective and transparent means of reaching conclusions regarding studies that produce statistical results. Critics have not unified around an alternative. Other forms of reporting confidence or uncertainty could probably grow in popularity. One strong critic of significance testing suggested a list of reporting alternatives: effect sizes for importance, prediction intervals for confidence, replications and extensions for replicability, meta-analyses for generality. None of these suggested alternatives produces a conclusion/decision. Lehmann said that hypothesis testing theory can be presented in terms of conclusions/decisions, probabilities, or confidence intervals. "The distinction between the ... approaches is largely one of reporting and interpretation."

On one "alternative" there is no disagreement: Fisher himself said, "In relation to the test of significance, we may say that a phenomenon is experimentally demonstrable when we know how to conduct an experiment which will rarely fail to give us a statistically significant result." Cohen, an influential critic of significance testing, concurred, "... don't look for a magic alternative to NHST *[null hypothesis significance testing]* ... It doesn't exist." "... given the problems of statistical induction, we must finally rely, as have the older sciences, on replication." The "alternative" to significance testing is repeated testing. The easiest way to decrease statistical uncertainty is by obtaining more data, whether by increased sample size or by repeated tests. Nickerson claimed to have never seen the publication of a literally replicated experiment in psychology. An indirect approach to replication is meta-analysis.

Bayesian inference is one proposed alternative to significance testing. (Nickerson cited 10 sources suggesting it, including Rozeboom (1960)). For example, Bayesian parameter estimation can provide rich information about the data from which researchers can draw inferences, while using uncertain priors that exert only minimal influence on the results when enough data is available. Psychologist John K. Kruschke has suggested Bayesian estimation as an alternative for the *t*-test. Alternatively two competing models/hypothesis can be compared using Bayes factors. Bayesian methods could be criticized for requiring information that is seldom available in the cases where significance testing is most heavily used. Neither the prior probabilities nor the probability distribution of the test statistic under the alternative hypothesis are often available in the social sciences.

Advocates of a Bayesian approach sometimes claim that the goal of a researcher is most often to objectively assess the probability that a hypothesis is true based on the data they have collected. Neither Fisher's significance testing, nor Neyman–Pearson hypothesis testing can provide this information, and do not claim to. The probability a hypothesis is true can only be derived from use of Bayes' Theorem, which was unsatisfactory to both the Fisher and Neyman–Pearson camps due to the explicit use of subjectivity in the form of the prior probability. Fisher's strategy is to sidestep this with the p-value (an objective *index* based on the data alone) followed by *inductive inference*, while Neyman–Pearson devised their approach of *inductive behaviour*.

Philosophy

Hypothesis testing and philosophy intersect. Inferential statistics, which includes hypothesis testing, is applied probability. Both probability and its application are intertwined with philosophy. Philosopher David Hume wrote, "All knowledge degenerates into probability." Competing practical definitions of probability reflect philosophical differences. The most common application of hypothesis testing is in the scientific interpretation of experimental data, which is naturally studied by the philosophy of science.

Fisher and Neyman opposed the subjectivity of probability. Their views contributed to the objective definitions. The core of their historical disagreement was philosophical.

Many of the philosophical criticisms of hypothesis testing are discussed by statisticians in other contexts, particularly correlation does not imply causation and the design of experiments. Hypothesis testing is of continuing interest to philosophers.

Education

Statistics is increasingly being taught in schools with hypothesis testing being one of the elements taught. Many conclusions reported in the popular press (political opinion polls to medical studies) are based on statistics. An informed public should understand the limitations of statistical conclusions and many college fields of study require a course in statistics for the same reason. An introductory college statistics class places much emphasis on hypothesis testing – perhaps half of the course. Such fields as literature and divinity now include findings based on statistical analysis. An introductory statistics class teaches hypothesis testing as a cookbook process. Hypothesis testing is also taught at the postgraduate level. Statisticians learn how to create good statistical test procedures (like *z*, Student's *t*, *F* and chi-squared). Statistical hypothesis testing is considered a mature area within statistics, but a limited amount of development continues.

The cookbook method of teaching introductory statistics leaves no time for history, philosophy or controversy. Hypothesis testing has been taught as received unified method. Surveys showed that graduates of the class were filled with philosophical misconceptions (on all aspects of statistical inference) that persisted among instructors. While the problem was addressed more than a decade ago, and calls for educational reform continue, students still graduate from statistics classes holding fundamental misconceptions about hypothesis testing. Ideas for improving the teaching of hypothesis testing include encouraging students to search for statistical errors in published papers, teaching the history of statistics and emphasizing the controversy in a generally dry subject.

Hypothesis Testing

Sampling distributions are useful in finding the confidence intervals for the population parameters in terms of the unbiased estimates obtained from the sample data. Based on the sample data if a statement is made on population parameter, for example the compressive strength of concrete is 2000 psi, it is necessary to prove the validity of this statement.

Hypotheses

Hypotheses are the statements made on the population parameters such as the average desired speed of the car drivers is 75 km/h. Procedure adopted in testing the statistical validity of this statement is known as hypothesis testing. Sample data is used to prove or disprove this statement. In the hypothesis testing it is necessary to make two statements about the parameter subjected to testing. These two are given below;

H_0: $\mu = 75$ km/h

H_1: $\mu > 75$ km/h

H_0 is called null hypothesis and the other one is known as the alternative hypothesis. The alternative hypothesis, shown above, is a one-sided alternative hypothesis. If H_1 is $\mu \neq 75$ km/hthen it is called as two-sided alternative hypothesis since the average speed can be greater or lesser than 75 km/h.

This testing is specifically useful in knowing the effectiveness of some changes made in the model or experimental setup, improvement in the controlling measures etc. In case of concrete compressive strength, if some admixtures are used to improve the strength, or the changes in the proportion of the aggregates, it is necessary to know the after effects of these changes.

As discussed before, in hypothesis testing sample data is crucial and this testing is carried out on the basis of the statistics calculated from this data. There is a chance to commit error while testing the hypothesis since the testing is done on the basis of sample data.

Types of Errors

Two types of errors are normally committed while testing the hypothesis using the sample data. If the null hypothesis is rejected when it is true, then the error committed is known as type I error. When the analyst has failed to reject the null hypothesis even when it is not true then type II error is committed.

Probability of committing type I error is nothing but the significance level used for the hypothesis testing. Committing type I error can be controlled by the analyst by choosing the appropriate significance level for the hypothesis testing. Significance level used in the hypothesis testing controls the critical region for the test statistic. Since committing type I error is in control of the analyst, *rejecting the null hypothesis would be always a strong conclusion.*

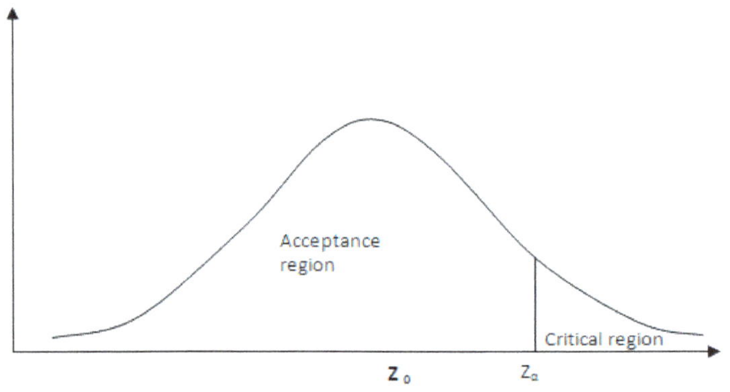

Acceptance and rejection regions for one-tailed (upper) testing

The figure above represents the acceptance and the rejection region regions in a one-tailed hypothesis testing.

For the example on desired speed of the car drivers and for the following hypotheses the acceptance and rejection regions for the null hypothesis are given in the figure. Assume that the sample size used in this testing is greater than 30 (even if population variance is unknown the mean of the sample data will follow normal distribution).

H_0: $\mu = 75$ km/h

H_1: $\mu > 75$ km/h

If the testing is conducted at 5% significance level the acceptance region is 95% of the area under normal curve and the rejection region is the remaining 5% area under the curve. Z_α (1.64 for 5% significance) is the critical value for the test statistic, Z_0, which is obtained from the sample data.

$$Z_0 = \frac{\overline{X} - \mu}{\sigma / \sqrt{n}}$$

If the null hypothesis is true Z_0 falls in the acceptance region. The probability of having the test statistic in the acceptance region is 1-α. If μ = 75 km/h is true the probability of test statistic falling in the critical region is α, and this may happen in unusual situations. This indicates the null hypothesis is false and shall reject it.

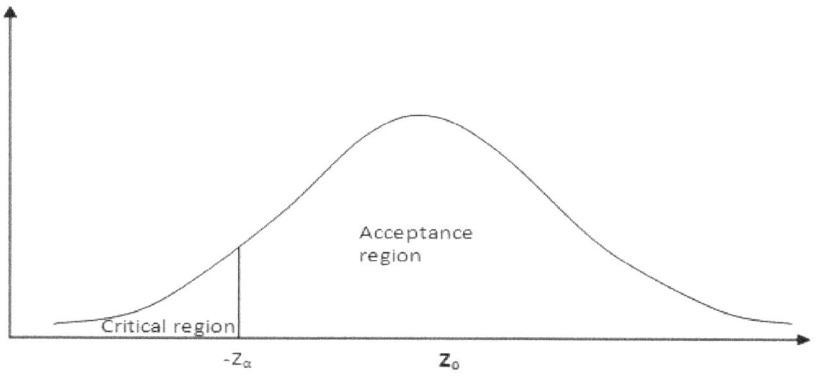

Acceptance and rejection regions for two-tailed testing

Instead, if the following hypotheses are used, the acceptance and rejection regions for the null hypothesis are given in the figure above.

H_0: μ = 75 km/h

H_1: μ ≠ 75 km/h

If the testing is conducted at 5% significance level the acceptance region is 95% of the area under normal curve and the rejection region is the remaining 5% area under the curve, but equally distributed on both the tails of the normal curve.

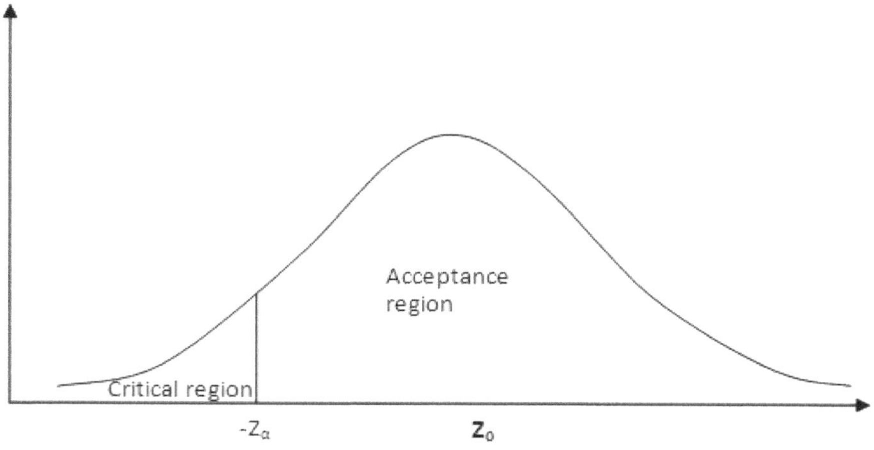

Acceptance and the rejection regions for one-tailed (lower) testing

Instead, if the following hypotheses are used, the acceptance and rejection regions for the null hypothesis are given in the figure above.

H_0: $\mu = 75$ km/h

H_1: $\mu < 75$ km/h

If the testing is conducted at 5% significance level the acceptance region is 95% of the area under normal curve and the rejection region is the remaining 5% area under the curve.

How to Formulate Hypotheses?

Hypotheses formulation requires understanding about the population data that is under consideration for testing. Hypotheses formulation related to the average speed of a particular vehicle type moving on a road section.

A road section has been upgraded and the analyst is optimistic about the increase in the average speed of the vehicles moving on that road section. In other words, the analyst is interested in proving that there is an increase in the average speed, compared to the speeds observed on the same road before up-gradation. As discussed earlier, rejecting the null hypothesis is a strong conclusion in hypotheses testing. On this basis, it can be said that whatever the hypothesis the analyst wants to reject must be taken as the null hypothesis. In the above example, null hypothesis may be "there is no significant change in the average speed of the vehicles even after upgrading the road section". Ideal outcome from this test should be 'reject null hypothesis', i.e. there is a significant change in the speeds of vehicles after up-gradation of the road section.

Some other examples are – If the problem is "when concrete properties are changed considerably when some admixtures are added" the null hypothesis may be no significant change in the concrete characteristics, - "In the trip generation modeling a particular variable is significantly contributing in explaining the variability of the trips generated", the null hypothesis may be that particular variable's contribution is nil, - "In a particular city modal share of public transport exceeds certain value", the null hypothesis may be the proportion of public transport users are less than that value.

Hypotheses Testing

Another crucial aspect of the hypotheses testing is deciding the significance level of the test, α. The significance of test is nothing but 'what is the probability of committing type I error (rejecting null hypothesis even when it is true)', given the sample data. Decision on its value alters the test result since the significance value decides the critical regions of the test. If the analyst goes for smaller α value, the critical regions for rejecting the null hypothesis would be smaller and rejection of null hypothesis may not be possible.

The procedure for hypotheses testing is given below;

1. Formulate the null hypothesis, H_0. In the hypotheses testing the decision is made on this statement. As discussed earlier this statement should be the one the analyst is interested in rejecting. The acceptance and the rejection regions shown in the figure are related to this statement.

2. Formulate the alternate hypothesis, H_1. This is what the analyst is interested in proving with the given sample data.

3. Identify and compute the test statistics from the sample data.

4. Identify the corresponding statistical distribution. This was discussed in detail in the previous lectures.

5. Assume significance level at which the test is to be carried out. Compute the confidence interval for the population parameter which is being tested. This interval is nothing but the acceptance region for the null hypothesis. Alternatively, the critical regions can also be identified based on the assumed significance level and the sampling distribution of the test statistic.

6. Test whether the statistic is falling in the acceptance region or not. If the test statistic is not falling in the acceptance region reject the null hypothesis. Rejecting null hypothesis implies the analyst is able to prove his statement on population parameter based on the given sample data.

Hypothesis Testing on Population Mean

Two situations may arise out of this, first one is when the population variance is known and the second situation is if the population variance is unknown.

Population Variance Known

The hypothesis on population mean is necessary when it is unknown. Hypothesis is nothing but assuming certain mean for population data. In hypothesis testing, using the sample data collected from the same population the truth in the assumption would be tested. The hypotheses are formulated as follows;

$$H_0 : \mu = \mu_0$$

$$H_1 : \mu \neq \mu_0 \, or \, H_1 : \mu > \mu_0 \, or \, H_1 : \mu < \mu_0$$

As discussed above another crucial aspect in hypothesis testing is the nature of the sampling distribution. If the Population is normal then the sampling statistic, i.e. $\bar{X}$, will also be normally distributed. Since the population variance σ^2 is known the test statistic for this hypothesis testing would be as follows;

$$Z_0 = \frac{\bar{X} - \mu}{\sigma / \sqrt{n}}$$

If the significance of the test is α then from the normal tables the values $-Z_{\alpha/2}$ and $Z_{\alpha/2}$ (for two tailed testing or Z_α for one tailed testing) can be obtained. If the value of Z_0 is falling in the critical region, reject the null hypothesis, if not falling in the critical region it can be said that the analyst has failed to reject the null hypothesis. Rejecting null hypothesis means the assumed population mean is different from the actual population mean at the significance level α. If the test is conducted at a different significance level the result of the hypothesis testing would be different.

Using the same sample data, the least significance level at which the null hypothesis can be rejected is known as p-value. If the test statistic follows standard normal distribution the p-values can be found as follows;

$$P = \begin{cases} 2[1-\Phi(|Z_0|)] \text{ for a two tailed test} \\ 1-\Phi(Z_0) \text{ for an upper tail test} \\ \Phi(Z_0) \text{ for a lower tail test} \end{cases}$$

$\Phi(Z_o)$ represents the area under the normal curve on the left side of Z_o.

Another Method for Testing the Hypotheses

Another way of testing the hypotheses is to find the confidence interval for the population parameter. If the assumed population mean is not falling in the confidence interval then the null hypothesis is rejected otherwise the conclusion is that the analyst has failed to reject the null hypothesis.

Problem

A company claims to fill its bags with 50 kg of cement (σ = 0.25 kg). A sample of 10 bags is taken, and found that on an average a bag weighs 49.75 kg. Use a 1% significance level to test if the data are consistent with this claim.

Answer:

Given data:
Standard deviation of the population data = 0.25 kg
Sample size = 10 bags
Sample mean = 49.75 kg
Significance level = 1%

Hypotheses formulation:
Null hypothesis: μ = 50 kg or μ ≥ 50 kg
Alternate hypothesis: μ< 50 kg

Test statistic = sample mean.

We know that the statistic follows normal distribution with mean μ and standard deviation $\sigma/\sqrt{n}$, then,

$$Z = \frac{\bar{X}-\mu}{\sigma/\sqrt{n}} = \frac{49.75-50}{0.25/\sqrt{10}} = -3.16$$

Z value corresponding to 1 % significance level (for one tailed testing) is -2.33.

Acceptance and the rejection regions:

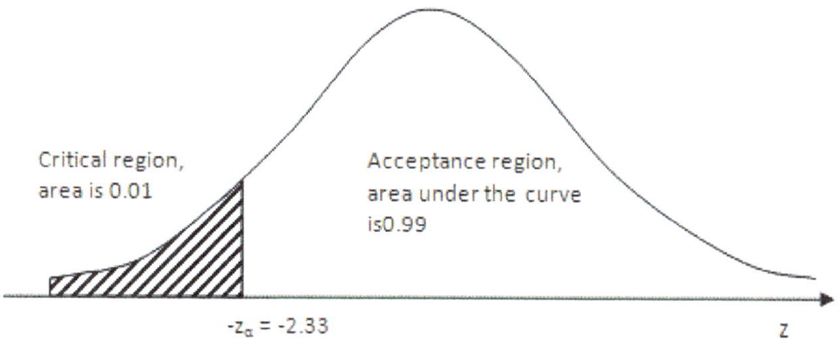

Acceptance and the critical regions for 99% upper confidence interval

Since the test statistic is falling in the critical region the conclusion is "reject the null hypothesis". The answer to the problem is data are not consistent with the company's claims.

Population Variance Unknown

When the population variance is unknown the unbiased estimator, sample variance, can be used in computing the test statistic. When the sample of size greater than or equals to 30 is used in the testing, the test statistic would be as follows;

$$Z_0 = \frac{\bar{X} - \mu}{S/\sqrt{n}}$$

The two-sided confidence interval for the population mean is,

$$\bar{X} - Z_{\alpha/2}\frac{s}{\sqrt{n}} \le \mu \le \bar{X} + Z_{\alpha/2}\frac{s}{\sqrt{n}}$$

For the hypotheses formulation shown below,

H_0: $\mu = \mu_0$

H_1: $\mu \neq \mu_0$

If μ_0 is falling in the confidence interval then test result is "failing to reject the null hypothesis", if not the result is "reject the null hypothesis".

When the sample is of size lesser than thirty the test statistic would be as given below.

$$t_0 = \frac{\bar{X} - \mu}{s/\sqrt{n}}$$

It follows t-distribution with n-1 degrees of freedom. .

Problem

The foundation for a building is designed to rest on 100 piles and the capacity of the pile is 80 tons. Nine test piles were driven at random and the failure loads were recorded. The results of the 9 tests are given below;

Pile no.	Capacity (Tons)
1	82
2	75
3	95
4	90
5	88
6	92
7	78
8	85
9	80

Based on these results, at 5% significance level, find out whether the foundation based on the piles should be accepted?

Answer

Given data:
Sample size = 9
Sample mean = 85 tons
Sample standard deviation = 6.76 tons
Significance level = 5%

Hypotheses formulation:
Null hypothesis: $\mu \le 80$ ton or $\mu = 80$ ton
Alternate hypothesis: $\mu > 80$ ton

Since the sample is small and the population variance is unknown, the statistic follows t – distribution with mean μ and standard deviation $s/\sqrt{n}$, then,

$$t = \frac{\bar{X} - \mu}{\sigma/\sqrt{n}} = \frac{85 - 80}{6.76/\sqrt{9}} = 2.22$$

t – Value corresponding to 5 % significance level (for one tailed testing) is 1.8595.

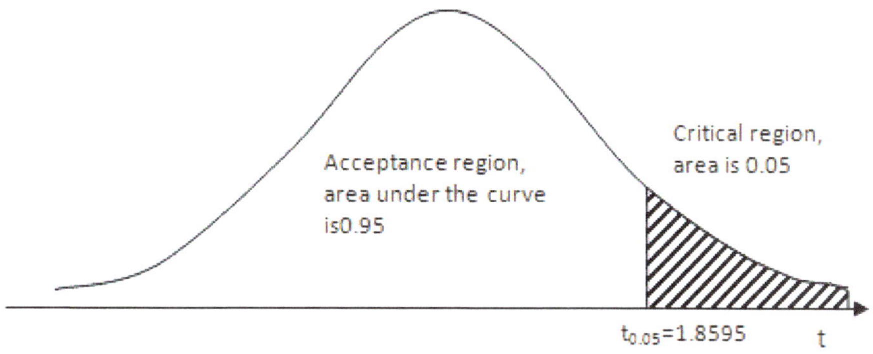

Acceptance and the critical regions for 95% lower confidence interval

Since the test statistic is falling in the critical region the conclusion is "reject the null hypothesis". The answer to the problem is "the foundation based on the piles can be accepted on the basis of test pile data".

Note: The hypotheses formulation in the above two problems may lead to some confusion. Motive in the hypotheses formulation, as discussed earlier, should be to reject the null hypothesis when it is true or commit type 1 error. But it may not be required to strictly follow this and depending on situation i.e., if some other information is available other than the data used in the hypotheses testing, type II error can also be committed.

Hypotheses Testing on the Means of two Population Data

When two populations of interest are to be compared using the unknown means and known variances the following procedure is to be adopted. If the populations are normally distributed and have the parameters μ_1, σ_1^2 and μ_2, σ_2^2, respectively, the comparison between μ_1 and μ_2 would be made based on two samples collected from the respective populations. Hypotheses on the two means could be as follows;

$H_0: \mu_1 = \mu_2$

$H_1: \mu_1 \neq \mu_2$

Or

$H_0: \mu_1 - \mu_2 = 0$

$H_1: \mu_1 - \mu_2 \neq 0$

Here the test statistic would be the difference in the means of both the samples, that is $\overline{X}_1 - \overline{X}_2$. If both the populations are normally distributed, this statistic is normally distributed with

parameters $\mu_1 - \mu_2$ and $\frac{\sigma_1^2}{n_1} + \frac{\sigma_2^2}{n_2}$. The transformed test statistic would be as follows;

$$Z_0 = \frac{\overline{X}_1 - \overline{X}_2}{\sqrt{\frac{\sigma_1^2}{n_1} + \frac{\sigma_2^2}{n_2}}}$$ with mean 0 and variance 1.

When the Variances are Unknown

As with the case of tests on the population mean when the variances are unknown the test statistic follows t-distribution. The following are the two sub cases that may arise out of this problem.

When the Variances of the Two Populations are not Equal

Since the variances of both the populations are unknown it can be assumed that either both of them are equal or not equal. In this case it is assumed that both the variances are not equal.

The sampling distribution of $\bar{X}_1 - \bar{X}_2$ would have mean, $\mu_1 - \mu_2$ and variance, $\frac{s_1^2}{n_1} + \frac{s_2^2}{n_2}$

$H_0 : \mu_1 = \mu_2$

$H_1 : \mu_1 \neq \mu_2$

For the above hypotheses,

$t_0 = \frac{\bar{X}_1 - \bar{X}_2}{\sqrt{\frac{s_1^2}{n_1} + \frac{s_2^2}{n_2}}}$ is t-distributed with the degrees of freedom as given below;

$$Degrees\ of\ freedom \left(\frac{\left(\frac{s_1^2}{n_1} - \frac{s_2^2}{n_2} \right)^2}{\left(\frac{s_1^2}{n_1} \right)^2 \Big/ n_1 + 1^{+} \left(\frac{s_1^2}{n_1} \right)^2 \Big/ n_1 + 1} - 2 \right).$$

When the Variances of the Two Populations are Equal

Important aspect of this would the common variance for both the populations since it is assumed that the variances are equal. Sample data taken from both the samples would be used in getting the variance of the pooled data.

Variance of the pooled data, $s_p^{\ 2} = \frac{(n_1 - 1)s_1^2 + (n_2 - 1)s_2^2}{n_1 + n_2 - 2}$

$H_0 : \mu_1 = \mu_2$

$H_1 : \mu_1 \neq \mu_2$

For the above hypotheses the test statistic is

$$t_0 = \frac{\bar{x}_1 - \bar{x}_2}{S_p \sqrt{\frac{1}{n_1} + \frac{1}{n_2}}}$$

which is t-distributed with $n_1 + n_2 - 2$ degrees of freedom.

Hypotheses Testing on the Variance

In this case also both the mean and the variance of the population are unknown. The hypotheses would be on the variance of the population data exceeding certain value, σ_0^2.

$H_0: \sigma^2 = \sigma_0^2$

$H_1: \sigma^2 \neq \sigma_0^2$

In this case also the formulation of hypotheses requires a thorough understanding of the problem in hand.

Test statistic in this case follows chi-square distribution with n-1 degrees of freedom and is as given below.

$$\chi_0^2 = \frac{(n-1)S^2}{\sigma_0^2}$$

Critical value for the χ_0^2 can be obtained from the standard tables available, given the degrees of freedom and the significance used for the testing. For a two tailed testing the null hypothesis would be rejected when

$$\chi_0^2 > \chi^2_{\alpha/2.n-1} \text{ and } \chi^2_{1-\alpha/2.n-1}$$

Otherwise it may be said that the analyst failed to reject the null hypothesis.

Hypotheses Testing on two Variances

When two independent populations are normally distributed and it is necessary to know whether the variability in both the populations is equal or not, the following procedure may be used. In this case also mean and the variances of both the populations are unknown. If the variances of both the populations are σ_1^2 and σ_2^2 respectively, the following are the typical hypotheses for this problem;

$H_0: \sigma_1^2 = \sigma_2^2$

$H_1: \sigma_1^2 \neq \sigma_2^2$

The test statistic, $F_0 = \frac{S_1^2}{S_2^2}$, follows F-distribution with $n_1 - 1$ and $n_2 - 1$ degrees of freedom. Based on the significance levels used in testing, from the standard F-tables, the critical regions can be decided.

Hypotheses Testing on Proportion

Testing on proportions of different types of data in the population has several applications in Civil Engineering. A simple application is finding out the proportion of public transport users among the trip makers of a particular area. It is reasonable to assume that the proportion of public transport users in the total trip makers follow binomial distribution. For hypotheses testing purposes the sampling distribution of proportion in the sample can be assumed to follow normal distribution with mean np_o and variance $np_o(1-p_o)$. This is true particularly for large nvalues.

H_o: $p = p_o$

H_1: $p \neq p_o$

The test statistic for the above hypotheses is as given below;

$$Z_0 = \frac{X - np_0}{\sqrt{np_0(1 - p_0)}}$$

follows standard normal distribution. Similar to the procedure discussed above, once the significance level is fixed the critical regions can be found out. Reject null hypothesis if the test statistic is falling in the critical regions (for two tailed testing).

Hypotheses Testing for Goodness of Fit

When the population distribution function is known to follow certain probability distribution this test is used for knowing the goodness-of-fit. Here also a sample data is to be collected from the population data and the frequency analysis has to be carried out on the data. The total data are divided into k classes, mostly depending upon the range of the observed data and any particular class must not have less than 8 observations. It is also necessary to calculate the parameters of the assumed probability distribution from the sample data. For example, if the time headways observed at a particular road section are assumed to follow negative exponential distribution, the mean headway is to be calculated from the observed data.

With the help of the assumed distribution and the parameters calculated from the sample data theoretical frequencies are to be generated. If O_i is the observed frequency and the E_i is the expected frequency then the following is the test statistic;

$$\chi_0^2 = \sum_{i=1}^{k} \frac{(O_i - E_i)^2}{E_i}$$

The test statistic follows chi-square distribution with k-p-1 degrees of freedom. p is the number of parameters in the assumed distribution and the 1 is used since the total frequency has been used in getting the theoretical frequency. From the assumed significance level and the degrees of freedom the critical regions are determined.

In this case the null hypothesis is 'the assumed distribution fits to the population data at the given significance level'. If the test statistic falls in the critical region reject the null hypothesis otherwise it may be said that the analyst has failed to see any difference between the theoretical and observed frequencies hence failed to reject the null hypothesis.

As can be seen from the formula for the chi-square observed value, it always takes positive values and no information is available on the location of the differences between the observed and the theoretical frequencies.

Example

It is hypothesized that the headway data obtained from high traffic flow conditions follow Erlang distribution. To know whether it is really true or not 352 observations were obtained. The mean headway is found to be 2.18 sec and the variance is 1.56 sec². The parameters for the Erlang distribution are 3 (shape parameter) and 1.376 (scale parameter). Obtained theoretical frequencies are tabulated in the table below. Theoretical frequencies are obtained using the following approximation for cumulative distribution function for the Erlang distribution.

$$p(H \leq h) = 1 - e^{\lambda * h}\left\{\sum_{i=1}^{k-1}\frac{(\lambda h)^i}{i!}\right\}$$

k is the shape parameter (only integer values) and λ is the scale parameter.

Chi-square calculated is 20.61. If the null hypothesis is "observed headway data follows Erlang distribution" and the alternate hypothesis is "data won't fit into Erlang distribution" at 1% significance level the conclusion of the hypotheses testing is "failed to reject the null hypothesis". This conclusion is based on the theoretical and calculated chi-square values. Chi-square calculated, 20.61, is less than the chi-square theoretical value, 21. 67 (at 9 degrees of freedom and 1% significance level), results in the above decision.

Table: Sample calculations for Chi-square goodness of fit test for the headway data (assumed to follow Erlang distribution)

Lower limit (h_1)	Upper Limit (h_2)	Observed freq.	$p(H \leq h_2)$	$p(h_1 < H \leq h_2)$	Expected. Freq.	$\frac{(Obs.-Expected)^2}{Expected}$
0	1	35	0.166	0.166	58.4	9.396
1	2	161	0.53	0.364	128.1	8.433
2	3	86	0.789	0.259	91.2	0.292
3	4	46	0.917	0.128	45.1	0.019
4	5	13	0.97	0.053	18.7	1.71
5	6	7	0.9898	0.0198	6.9	0.00
6	7	2	0.9967	0.0069	2.4	0.08
7	8	1	0.9989	0.0022	0.8	0.68
8	9	0	0.9997	0.0008	0.3	
9	10	1	0.9999	0.0002	0.1	

Other Tests for Goodness-of-Fit

Most of the times, it is assumed that the population data follow certain distribution (in majority of the cases it is assumed to follow normal distribution). There are also problems where the distributional assumption cannot be made on the population data. In this scenario, some non-parametric tests are available for testing the goodness-of-fit of the observed and the theoretical densities.

Kolmogorov – Smirnov Goodness of fit Test

This is a non-parametric test and applicable to the continuous random variables. In this case it is not required to make the distributional assumption and testing is carried out on the cumulative distribution functions.

In this case the test statistic is the maximum vertical separation between the theoretical and the observed cumulative distribution functions. If $x_1, x_2, \ldots, x_n$ are the order statistics of a continuous random variable X arranged in the ascending order, the cumulative distribution function is as follows;

$F_n(x) = 0$ for $X < x_1$

$= k/n$ for $x_k \le X \le x_{k+1}$; $k = 1,2,\ldots,n-2.$

$= 1$ for $X \ge x_n$

Critical values at different significance levels are different and depend on the following situations.

1. Theoretical distribution parameters are not estimated from the sample data (some other sample might have been used in getting the parameter values
2. Parameters are estimated from the same sample data.

For the first case, for large samples (sample size > 40) the critical values are as follows (Miller, L.H. 1956);

$$D_{n,0.10} = 1.22/\sqrt{n},\ D_{n,0.05} = 1.36/\sqrt{n},\ D_{n,0.02} = 1.52/\sqrt{n},\ D_{n,0.01} = 1.63/\sqrt{n}$$

For the second case, values are provided by Dallal and Wilkinson (1986).

For the headway problem the results of the K-S test are as follows;

The CDF of the observed data and the Erlang distribution are shown in the figure below. The Maximum vertical separation is 0.074. Null hypothesis is "data fit to the Erlang distribution" and the alternate hypothesis is "data differ from the Erlang distribution". Since the parameters are estimated from the same data the critical value of the statistic is obtained from the Table provided in Dallal and Wilkinson (1986). At 1 % significance level the critical value is 0.0603. Since this value is less than the test statistic the conclusion is "reject the null hypothesis".

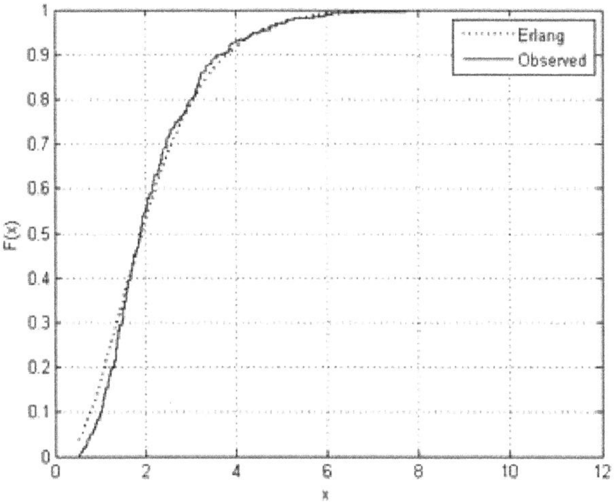

Observed and the Erlang cumulative distribution functions for the time headway data

Anderson – Darling Goodness-of-fit Test

This test is useful when some values in the data are quite smaller or larger compared to the sample mean, in other words data has several outliers.

Example Problems

The following problems have been formulated from the travel data collected from the elderly people (more than 60 years old) of Shimla, Himachal Pradesh. In these problems, travel related aspects of female and male travelers, like average walking time, average number of trips in terms of mean and the variability, have been compared.

Average Walking Time:

	Average Walking Time (in min) to Shimla downtown	**Standard Deviation**
Men	32.61	12.76
Women	35.85	14.78
Total Sample	33.95	13.71

Hypotheses Testing for Variance Inequality between avg. Walking Time of Elderly Male and Female Trip Makers

This problem deals with the testing of the variability in the walking times of elderly male and female trip makers in the Shimla town. Here, the two populations are the elderly male and female trip makers. Sample size collected from the male population is 297 and from the female population it is 210.

Variable 1 = Walking time for men

Variable 2 = Walking time for women

Hypotheses formulation for this problem is as follows;

$H_0: \sigma_1^2 = \sigma_2^2$

$H_1: \sigma_1^2 \neq \sigma_2^2$

Sample means are 32.6 and 35.9 minutes for male and female travelers, respectively. Sample variances are 162.8 and 218.4 for both the variables respectively.

As discussed earlier the test statistic in this case follows F- distribution.

$$F_0 = \frac{S_1^2}{S_2^2}$$

Test on the Variances		
	Variable 1	*Variable 2*
Mean	32.6	35.9
Variance	162.8	218.4
Observations	297	210
df	296	209
F calculated	0.745	
P(F<=f) one-tail	0.0102	
F Critical one-tail	0.812	

p-value of 0.01091484 (< 0.05) indicates that the variances are not equal so we will go for t-test for unequal variances for avg. walking time for men and women.

t-test for avg. walking time of men and women

Variable 1 = Walking time for men

Variable 2 = Walking time for women

t-Test: Two-Sample Assuming Unequal Variances		
	Variable 1	*Variable 2*
Mean	32.60606061	35.85714286
Variance	162.8138821	218.4005468
Observations	297	210
Hypothesized Mean Difference	0	
df	407	
t Stat	-2.579738612	
P(T<=t) one-tail	0.005118641	
t Critical one-tail	1.64860612	
P(T<=t) two-tail	0.010237281	
t Critical two-tail	1.965809661	

As t stat > 1.96 hence the difference between avg. walking time of men and women is significant. (Rejecting null hypothesis at 95% confidence level).

Problem for Chi-Square Goodness of Fit Test

In this example the headway data collected on a road near Guwahati has been fitted to the negative exponential distribution and theoretical headway frequencies are obtained from the negative exponential distribution.

Table: Calculations involved in testing the closeness of the observed data with the assumed probability distribution

Class	Head-way, t	Observed fre-qunecy	Obs. Freq. h>t	Relative freq. (h>t)	Theor. Rel. freq.	Theor. Freq. (h>t)	Theor. Freq.	$\frac{(O_i - E_i)^2}{E_i}$
0 to 2	0.00	47	522.00	100.00	100.00	522.00	159	77.98
2 to 4	2.00	125	475.18	91.03	69.59	363.28	110	2.05
4 to 6	4.00	128	398.18	76.28	48.43	252.82	77	35.58
6 to 8	6.00	67	270.18	51.76	33.71	175.94	53	3.70
8 to 10	8.00	49	203.18	38.92	23.46	122.45	37	3.89
10 to 12	10.00	32	154.18	29.54	16.32	85.21	26	1.96
12 to 14	12.00	11	122.18	23.41	11.36	59.30	18	2.72
14 to 16	14.00	11	111.18	21.30	7.91	41.27	13	0.08
16 to 18	16.00	13	100.18	19.19	5.50	28.72	9	3.13
18 to 20	18.00	7	87.18	16.70	3.83	19.99	6	0.17
20 to 22	20.00	9	80.18	15.36	2.66	13.91	14	1.23
								132.48

The statistic resulting from the last column of the table above follows chi-square distribution. Hypotheses formulation for this problem is given below;

H_0: Headway data follows negative exponential distribution

H_1: Headway data doesn't follow the negative exponential distribution

Chi-square calculated is 132.48 and the critical Chi-square value, for 11-1-1 = 9 degrees of freedom (1 for the mean headway and the other one for cumulative frequency) at 5 % significance level, is 3.33. Since the critical value is less than the calculated value the decision is "reject the null hypothesis that the headway data follows negative exponential distribution".

References

- Fisher, R (1955). "Statistical Methods and Scientific Induction" (PDF). Journal of the Royal Statistical Society, Series B. 17 (1): 69–78
- Lehmann, E. L.; Romano, Joseph P. (2005). Testing Statistical Hypotheses (3E ed.). New York: Springer. ISBN 0-387-98864-5
- C. S. Peirce (August 1878). "Illustrations of the Logic of Science VI: Deduction, Induction, and Hypothesis". Popular Science Monthly. 13. Retrieved March 30, 2012
- Gigerenzer, G (November 2004). "Mindless statistics". The Journal of Socio-Economics. 33 (5): 587–606. doi:10.1016/j.socec.2004.09.033
- Hinkelmann, Klaus and Kempthorne, Oscar (2008). Design and Analysis of Experiments. I and II (Second ed.). Wiley. ISBN 978-0-470-38551-7
- Cornfield, Jerome (1976). "Recent Methodological Contributions to Clinical Trials" (PDF). American Journal of Epidemiology. 104 (4): 408–421

- Bakan, David (1966). "The test of significance in psychological research". Psychological Bulletin. 66 (6): 423–437. doi:10.1037/h0020412
- Montgomery, Douglas (2009). Design and analysis of experiments. Hoboken, N.J.: Wiley. ISBN 978-0-470-12866-4
- Begg, Colin B.; Berlin, Jesse A. (1988). "Publication bias: a problem in interpreting medical data". Journal of the Royal Statistical Society, Series A: 419–463
- Neyman, J; Pearson, E. S. (January 1, 1933). "On the Problem of the most Efficient Tests of Statistical Hypotheses". Philosophical Transactions of the Royal Society A. 231 (694–706): 289–337. doi:10.1098/rsta.1933.0009
- Richard J. Larsen; Donna Fox Stroup (1976). Statistics in the Real World: a book of examples. Macmillan. ISBN 978-0023677205
- Journal of Articles in Support of the Null Hypothesis website: JASNH homepage. Volume 1 number 1 was published in 2002, and all articles are on psychology-related subjects
- Goodman, S N (June 15, 1999). "Toward evidence-based medical statistics. 1: The P Value Fallacy". Ann Intern Med. 130 (12): 995–1004. PMID 10383371. doi:10.7326/0003-4819-130-12-199906150-00008
- Moore, David (2003). Introduction to the Practice of Statistics. New York: W.H. Freeman and Co. p. 426. ISBN 9780716796572
- Kruschke, J K (July 9, 2012). "Bayesian Estimation Supersedes the T Test". Journal of Experimental Psychology: General. 142: 573–603. doi:10.1037/a0029146
- Neyman, Jerzy (1967). "RA Fisher (1890—1962): An Appreciation.". Science. 156.3781: 1456–1460. doi:10.1126/science.156.3781.1456
- Jaynes, E. T. (2007). Probability theory : the logic of science (5. print. ed.). Cambridge [u.a.]: Cambridge Univ. Press. ISBN 978-0-521-59271-0
- Mayo, D. G.; Spanos, A. (2006). "Severe Testing as a Basic Concept in a Neyman–Pearson Philosophy of Induction". The British Journal for the Philosophy of Science. 57 (2): 323–357. doi:10.1093/bjps/axl003
- Loftus, G R (1991). "On the Tyranny of Hypothesis Testing in the Social Sciences" (PDF). Contemporary Psychology. 36 (2): 102–105. doi:10.1037/029395
- Stigler, Stephen M. (1986). The History of Statistics: The Measurement of Uncertainty before 1900. Cambridge, Mass: Belknap Press of Harvard University Press. p. 134. ISBN 0-674-40340-1
- Meehl, P (1990). "Appraising and Amending Theories: The Strategy of Lakatosian Defense and Two Principles That Warrant It" (PDF). Psychological Inquiry. 1 (2): 108–141. doi:10.1207/s15327965pli0102_1

Permissions

All chapters in this book are published with permission under the Creative Commons Attribution Share Alike License or equivalent. Every chapter published in this book has been scrutinized by our experts. Their significance has been extensively debated. The topics covered herein carry significant information for a comprehensive understanding. They may even be implemented as practical applications or may be referred to as a beginning point for further studies.

We would like to thank the editorial team for lending their expertise to make the book truly unique. They have played a crucial role in the development of this book. Without their invaluable contributions this book wouldn't have been possible. They have made vital efforts to compile up to date information on the varied aspects of this subject to make this book a valuable addition to the collection of many professionals and students.

This book was conceptualized with the vision of imparting up-to-date and integrated information in this field. To ensure the same, a matchless editorial board was set up. Every individual on the board went through rigorous rounds of assessment to prove their worth. After which they invested a large part of their time researching and compiling the most relevant data for our readers.

The editorial board has been involved in producing this book since its inception. They have spent rigorous hours researching and exploring the diverse topics which have resulted in the successful publishing of this book. They have passed on their knowledge of decades through this book. To expedite this challenging task, the publisher supported the team at every step. A small team of assistant editors was also appointed to further simplify the editing procedure and attain best results for the readers.

Apart from the editorial board, the designing team has also invested a significant amount of their time in understanding the subject and creating the most relevant covers. They scrutinized every image to scout for the most suitable representation of the subject and create an appropriate cover for the book.

The publishing team has been an ardent support to the editorial, designing and production team. Their endless efforts to recruit the best for this project, has resulted in the accomplishment of this book. They are a veteran in the field of academics and their pool of knowledge is as vast as their experience in printing. Their expertise and guidance has proved useful at every step. Their uncompromising quality standards have made this book an exceptional effort. Their encouragement from time to time has been an inspiration for everyone.

The publisher and the editorial board hope that this book will prove to be a valuable piece of knowledge for students, practitioners and scholars across the globe.

Index